Can Machines Have Free Will? The Concept of Free Will in Relation to the Psychophysical Problem

Studies in Philosophy, History of Ideas and Modern Societies
Edited by Jan Hartman

Volume 26

Krzysztof Krenc

Can Machines Have Free Will? The Concept of Free Will in Relation to the Psychophysical Problem

PETER LANG

Berlin - Bruxelles - Chennai - Lausanne - New York - Oxford

Bibliographic Information published by the Deutsche Nationalbibliothek
The Deutsche Nationalbibliothek lists this publication in the Deutsche Nationalbibliografie; detailed bibliographic data is available in the internet at http://dnb.d-nb.de.

Library of Congress Cataloging-in-Publication Data
A CIP catalog record for this book has been applied for at the Library of Congress.

The cover image courtesy of Benjamin ben Chaim.

This publication was financially supported by University of Lodz, Poland.

ISSN 2191-1878
ISBN 978-3-631-90899-0 (Print)
E-ISBN 978-3-631-91076-4 (E-PDF)
E-ISBN 978-3-631-91077-1 (E-PUB)
DOI 10.3726/b21329

© 2023 Peter Lang Group AG, Lausanne
Published by Peter Lang GmbH, Berlin, Deutschland

info@peterlang.com - www.peterlang.com

All rights reserved.

All parts of this publication are protected by copyright. Any utilisation outside the strict limits of the copyright law, without the permission of the publisher, is forbidden and liable to prosecution. This applies in particular to reproductions, translations, microfilming, and storage and processing in electronic retrieval systems.

This publication has been peer reviewed.

Contents

Introduction 7

Chapter 1: Mainstream contemporary solutions to the so-called "free will problem" 9

Introduction 9
The problem of free will 9
Possible relations between free will and determinism 12
Incompatibilism 14
Meaning of "could have done otherwise" 15
Argument for incompatibilism: The consequence argument 16
Free will and the consequence argument 21
Compatibilism 29
Arguments for compatibilism 30
Discussion of arguments for compatibilism 34
Summary 40

Chapter 2: Free will and dualism 45

Introduction 45
Cartesian dualism 46
Arguments for Cartesian dualism 48
Freedom of will 51
Problems of interactionism in Cartesian dualism 56
Problems of Cartesian dualism with respect to free will problem 66
Other dualist solutions 68
Summary 75

Chapter 3: Free will and materialism 77
Introduction 77
Biological naturalism 78
Anomalous monism 87
Eliminative materialism 91
Summary 111

Chapter 4: Free will and transcendental idealism 113
Introduction 113
Theoretical problems with eliminative materialism 113
Transcendental idealism and free will 121
Things in themselves, appearances and eliminative materialism 130
Kant and free will in relation to compatibilism and incompatibilism 138
Kantianism and mysterianism 139
Summary 140

Conclusion 143

Bibliography 145

Introduction

The goal of this book is to analyse the notion of free will from the perspectives of various stances in the philosophy of mind. It employs an approach quite different than the more standard one, in which philosophers try to answer the question "do we have free will?" directly. Therefore, the first chapter contain an analysis of how this kind of problems are usually tackled, with the main focus on the contemporary compatibilism/incompatibilism debate – the question "is free will compatible with the determinism?" Showing that the answer to this question means something different for compatibilists and incompatibilists (*pace* van Inwagen) and the fact that it also depends on their assumptions about the mind (sometimes not explicitly stated), provides a rationale for the approach chosen in this book.

The subsequent chapters contain discussions about the relation between free will and dualism, materialism and transcendental idealism. The considerations about free will in relation to dualism serve as a background for the two final chapters, which contain the most important conclusions of the book. In the third chapter I discuss the relation between free will and materialism, with a focus on whether a particular kind of materialism called eliminative materialism gives any room to talk about free will. To do that, I go more in depth into the analysis of how artificial neural networks process information. Given that neural networks are commonly employed by materialistic philosophers as a model of the human mind, the analysis of the notion of free will in this context will give a better understanding of whether it is a sensible approach to understanding the human mind. In the last chapter, devoted to transcendental idealism, I present a solution to the free will problem (and Kant's third antinomy of reason) that incorporates eliminative materialism on the empirical level.

I believe the main value of the inquiry presented in this book is in the chosen context for the considerations of the idea of free will. Contemporary philosophy is mostly naturalistic and it is clear that naturalistic philosophers want to retain either the notion of free will or at least the notion of moral responsibility, but it is questionable whether they can achieve it. Also the discussions about "thinking machines" are contemporarily as prevalent as never before, not only among researchers and philosophers. Some are worried that artificial intelligence systems will destroy humanity and take over the Earth, while others wonder whether they should have moral rights similar to human beings. Free will has to be in the centre of this kind of considerations and I believe that the results presented in this book strongly support the idea that the difference between humans and mere machines is not just quantitative.

Chapter 1: Mainstream contemporary solutions to the so-called "free will problem"

Introduction

Issues that this thesis is about are denoted with a plethora of terms or phrases. "Problem of free will," "compatibility of determinism and moral responsibility," "elbow room" – just to mention a few. All of them are ambiguous, mainly because of the usage of ambiguous words. They also yield a plethora of solutions which are sometimes clearly contradictory, when the same words are used in different meanings, and sometimes coincidentally congruent. Therefore it is not enough to state, "I believe elbow room in human actions is not an illusion and here are my arguments for that," but a crucial part of a solution is an explanation what is meant by the phrase "elbow room." The aim of this chapter is to show different ways in which philosophers think about the problem of free will and other related ones and how they try to tackle them. On one hand it is important to have a good grasp of the mentioned solutions in order not to reinvent the wheel. On the other this knowledge let us show why those solutions are not satisfactory or even more – why they are meaningless when we consider them together as a solution to the same problem, as I will try to show. Therefore this chapter serves as the basis for understanding the rationale for presenting the argument for human autonomy present in the subsequent chapters.

The problem of free will

Let us start with a short introduction of the problem itself. Even though there is *some* philosophical problem that needs to be tackled, I have a strong conviction that the free will problem, as it is at length presented in the contemporary philosophical literature, is not possible to be tackled anymore. We can say that the so-called "problem of free will" starts with a problem of characterizing what "the problem of free will" is. Of course it is not distinctive of the free will problem – it is not hard to point out other issues like this in philosophy. But the problem of free will is nevertheless somehow special in this regard.

In its simplest form it can be characterized this way: "if determinism is true, how is free will possible?" If we look at this simple description even from its form we can get an initial understanding of what kind of a problem there is to face. Even without knowing the meaning of the terms "determinism" and "free will" we can say that there is tension between what they denote – if determinism is true, free will is questionable and the other way around. If I say, "how is it possible that the street is dry when it rained a moment ago?" I express a conviction that there must be other factors that came into play because rain and dry streets are not compatible with each other. This remark seems to be trivial, but as it will be shown later in case of the problem of free will it is not clear the things in question are incompatible with each other.

What exactly are the things in question? Even though probably everybody has some initial thoughts on what determinism and free will are, we have a lot of freedom in choosing what exactly we mean when we tackle the free will problem, and it is one of the things that cause a lot of philosophical troubles.

But let us try to start with as general definitions as possible. Among the two terms, it is probably easier to define what determinism is. A commonsense definition could be "everything that happens in the universe has something in the previous state of the universe as its cause." Although it might resemble common intuitions, again, it introduces a lot of freedom in interpreting its meaning. In philosophical literature determinism is commonly defined as "the thesis that the past and the laws of nature together determine, at every moment, a unique future" (Van Inwagen, 2017, 151) which makes a concise and at least seemingly clear definition.

On the other hand it is much harder to define "free will." I do not think I am able to provide a common-sense definition, so I will focus only on how philosophers describe free will, for now in the most general way. A common concept of free will used in philosophical literature is an idea that an agent possesses free will when at some point in time when the agent does something, he/she can also do something else or at least refrain from doing what he/she does. It is often expressed in the past tense: an agent acted freely while doing something when he could have also done otherwise. An idea that an agent could have done otherwise is described by Moore in his *Ethics* as an essential part of his theory of morals (Moore, 2005, 103), as it is needed for meaningfulness of moral deeds. This is also what Fieser calls "genuine free will[1]" (2009, 139). It is of course debatable whether it is the best possible definition (if a definition can be valued at all) and certainly it is also not clear that free will as defined is needed for meaningfulness of moral deeds (but, of course, here we would need to define "moral deeds"). But because this idea is so commonly discussed in literature (e.g. by Dennett (2003), Frankfurt (1969), Van Inwagen (2017), to name a few philosophers) that it makes it at least a good starting point. It is also this very idea, I believe, that make the contemporary discussion on free will not only hard and frustrating, but also futile or a Sisyphean task. These accusations may sound strong, but as I will try to show in this chapter, to be able to meaningfully talk about issues that are essential to the free will problem, we need to realize this.

But first things first, what is then the problem of free will if we consider the definitions outlined above? At first glance, it seems obvious: if determinism is true, then at each point of time there is only one possible future (it is actually a necessary future). So, any agent at any point in time, could not have chosen otherwise than they did, assuming that determinism is true. But, as expected, settling up the free will problem this way is just the beginning, not the end. Otherwise we could finish with the above conclusion and either wait for the natural sciences to give us information whether determinism is true in our universe or construct a philosophical argument for or against determinism, if it is a philosophical matter. But the presented problem is not a problem about the truth of determinism – it is a problem regarding the relation between free will and determinism, no matter whether determinism is true or not. In other words, someone who believes that the universe is not deterministic could still support an idea that the truth of determinism does not have to entail the denial of existence of free will.

---

1 "Genuine Free Will: for at least some actions, a person has the ability to have done otherwise."

Possible relations between free will and determinism

After these important preliminary remarks on the problem of free will let us stop keeping the issue in question on such a high level and get into details. There are of course two possible relations between free will and determinism regarding their compatibility: free will is either compatible with determinism or not. In philosophical literature philosophers who believe that free will is compatible with determinism are called "compatibilists" and their opponents are, not surprisingly, "incompatibilists." If free will is compatible with determinism it is possible to go even a step further and say that free will entails determinism. But it is not a necessary condition for compatibilism – it is possible to believe that free will is compatible with both determinism and indeterminism (where indeterminism is just a denial of determinism[2]) and this stance is sometimes called "supercompatibilism." On the other hand, if free will is incompatible with determinism it either has to be compatible with indeterminism or it does not exist at all because it is claimed to be incompatible with both determinism and indeterminism. Philosophers who believe that free will is incompatible with both determinism and indeterminism are called "hard incompatibilists," whereas philosophers that believe that free will is compatible only with indeterminism are called "libertarians." Philosophers that believe it is incompatible with determinism and believe that determinism is true are called "hard determinists." I believe it has already become hard to track all these different stances regarding the relation between free will and determinism, so let us try to list them using compatibility of free will and determinism as the main dividing criterion:

---

2 This "just" may be misleading as it is not entirely obvious what is negated by "in" in "indeterminism." If determinism is defined as "the thesis that the past and the laws of nature together determine, at every moment, a unique future," then its negation may refer to different parts of this definition. What I want to negate is the phrase "at every moment" – if indeterminism is true, then there was or there will be at least one moment in time when the past and the laws of nature do not determine a unique future. It is clear that this definition allows for very many ways in which events in our universe can be determined. What is important is that negation of determinism does not mean that all events do not have their efficient causes.

1. Compatibilism (sometimes called also "soft determinism"): free will is compatible with determinism.
   - Supercompatibilism: free will is compatible with both determinism and indeterminism.
   - Free will entails determinism (no name for this stance).
2. Incompatibilism: free will is not compatible with determinism.
   - Hard determinism: free will is incompatible with determinism and determinism is true.
   - Hard incompatibilism: free will is incompatible with both determinism and indeterminism.
   - Libertarianism: free will entails indeterminism.

So, if the above schema is correct, there are two main stances regarding free will and each of them has different flavours. Before discussing arguments for and against each of these stances it is important to at least mention an issue that is not only related to the free will problem, but that probably is the very thing that triggers the problem: the relation between moral responsibility, or responsibility in general, free will and determinism. Of course, this relation is even more complex than the previous one as it involves three parties. So for example it is possible to believe that moral responsibility is compatible with determinism even though we cannot say the same about free will. Also, there are philosophers that believe we need free will in order to preserve moral responsibility and therefore, by definition, moral responsibility entails free will. It is also debatable whether moral responsibility is the only responsibility we should care about or rather it is responsibility for every choice we make. As Searle wrote:

> Why the fuss about criminal and moral responsibility if every time you raised your arm, drank a beer, got married, joined the Communist Party, chose chocolate over vanilla ice cream, enrolled in a university course, decided not to commit suicide, or did more or less anything at all, you did so under a false presupposition? (2007b, 70)

Although it is possible to imagine "why the fuss" about criminal and moral responsibility, examples given by Searle are indeed interesting examples of non-moral choices to which determinism may pose a threat. Although it is harder to believe in importance of choosing one ice cream taste over another, many people do care deeply about with whom they are getting married, what political beliefs they have or which university course they enrol in, often losing an opportunity to choose another. Moral deeds are often considered more important probably because of how tremendous consequences would be, if they were ruled out, with putting into question the whole criminal justice system as a notable example. But nevertheless we can clearly see that the relation between responsibility, free will and determinism is extremely complicated and considering all possible relations between these three parties would rather blur the picture instead of bringing understanding. So in this chapter I will only refer to the idea of responsibility whenever it is necessary.

That being said let us discuss various stances regarding the relation between free will and determinism.

Incompatibilism

I believe compatibilism as a stance is a response to incompatibilism (not the other way around) and therefore it is more instructive to start with arguments driving incompatibilist convictions. As I will also argue, it is incompatibilism that gives rise to the free will problem as described above.

Let us start with the second thing. How did it actually happen that we talk about the "free will problem"? What could the first philosopher, maybe even pre-Pre-Socratic, who considered the issues we are occupied with, think that led him/her to his/her inquiry? Of course we can track history of philosophy and find the first philosopher that talked about free will. According to Long and Sedley it was Epicurus and they describe his problem this way:

> Epicurus' problem is this: if it has been necessary all along that we should act as we do, it cannot be up to us, with the result that we would not be morally responsible for our actions at all. (1987, 107)

It clearly states an incompatibilist stance, although without any visible argument why it cannot be up to us what we do if it has been necessary all along. When reading this short passage we have to assume that the meaning of "necessary" and "up to us" in this sentence exclude each other. But we do not really have to consider a particular philosopher that thought about free will for the first time in history of philosophy to be able to see the problem's origin. For a philosophical problem of this sort to emerge, some kind of incompatibility (at least alleged) is just needed. And we can imagine the first-person ever to think about the free will problem as somebody perceiving causal determinism in the world around them and realizing that if they make a decision, they may be determined to do what they do in the same way as the perceived world. Using Epicurus' phrase, their action is not really up to them.

But if it was so simple, then the first step to check whether free will is possible would be to establish whether determinism is true. If it were, we would know free will is not possible. But since there are multiple arguments stating why free will and determinism are incompatible that go much deeper than the simple Epicurus' expression of incompatibilism, we may expect the relation between free will and determinism is not that obvious. Let us now then look more closely at the idea of incompatibilism and arguments that drive it.

Meaning of "could have done otherwise"

Probably the clearest possible way to express what incompatibilism means is present in van Inwagen's *Critical Study of Dennett's Elbow Room* (2017). He makes there an important remark that "you could have done otherwise" can have two different meanings: "you were able to do otherwise" and "you might have done otherwise." The first is an expression of personal feature of an agent. The latter expresses modality. The difference is very well visible in one of the examples given by Van Inwagen (which he adapted from Austin). "You could have" in "You could have ruined me this morning, but you didn't. I owe you." clearly means "you were able," while in "You could have ruined me this morning. Warn me the next time you're going to pull a stunt like that one." means "you might have."

Now there is an important question about the relation between these two meanings of "you could have." It is quite clear that "you might have" does not entail "you were able" and the Austin's example depicts it very well. There are situations in which one might have done something while not being able to do it, like for example when tossing a coin: one may toss a coin in a way that it will turn heads and at the same time have no ability to make it so. It should be obvious that sometimes one might have done something and was able to do something – for example usually when one made a coffee they usually must have been able to do it and clearly might have done it (although we can probably think about edge cases when one was not able to make coffee and the lack of this ability did not prevent them from making coffee). Much more interesting is the question whether "you were able" entails "you might have." According to van Inwagen (2017, 52), the attitude towards this entailment defines the difference between incompatibilists and compatibilists. Incompatibilists believe that this entailment holds, whereas compatibilists deny it. So if determinism is true it is enough for an incompatibilist to use the law of contrapositive to make a conclusion that nobody could ever have done otherwise than they did.

Now, what does it mean that some people believe an entailment holds and some deny it? An entailment either holds or not and it is not a matter of personal opinion about it. But in this case we may say that the meaning of "could have done otherwise" is underdetermined and without further investigation it is not possible to determine which party is right. And the discussion between compatibilists and incompatibilists has at its core consideration of different situations that are supposed to show whether this entailment holds (even if it is not always expressed this way). Let us now discuss the main argument for the meaning of "could have done otherwise" that is used by incompatibilists.

Argument for incompatibilism: The consequence argument

Kane writes in *The Significance of Free Will*: "Free will (…) is the power of agents to be the ultimate creators (or originators) and sustainers of their own actions." (1998, 4). While this sentence may be easily attacked for not making clear what it means to be the ultimate creator of one's action (which Kane admits in the same book (1998, 5)), it is quite easy to show why determinism prevents the possibility of actions being ultimately originated by an agent, whatever it means. If determinism is true, every process in any brain has ultimate antecedent causes that are not part of this particular brain. Unless a brain was the first thing created in the universe, there must have been outer processes that shaped brains and how they work is a product of these processes. If determinism is true, then for each particular brain the way it is shaped depends on something that existed before this brain started to exist. If we were able to change the initial conditions of the universe, or the states of the universe before brains started to emerge during evolutionary processes, we could manipulate this process in the way that the same (numerically) brains would have different characteristics. Of course I do not want to say here that a brain in general, as part of the body, would be different, although it could also be the case. It is probably uncontroversial to say that how particular brains are formed is a function of the state of the universe before they were formed and the laws of nature – this is just a restating definition of the determinism. But if so, then if we had a power of changing the past, we would be able to change brains of particular people in the way that would influence their decisions without influencing themselves directly.

Of course even if we were able to manipulate the past, it would be extremely hard to shape it in the way that would make other people make concrete decisions that we want, especially if we wanted to make it happen to multiple people at the same time. But it does not matter anything for the conclusion – if determinism is true, our current decisions must be dependent on the state of the universe before our brains were created.

That is not to say that determinism implies fatalism, although these two ideas have a few common points and the relation between them is not that obvious. Fatalism is an idea that no matter what an agent does, he/she always ends up in the same place, obtains the same outcome of his/her action, etc. Dennett pictures fatalism with an example of a person being locked in a room (Dennett, 1984, 554). It does not matter if the person tries to open the door or not, he/she will remain inside, because the room is locked, and the fact that he/she is unaware of this does not change anything. Dennett calls this situation "local fatalism," but if we think of fatalism in general this way, it is easy to show how fatalism is different than determinism – it means that our actions will lead to the same results no matter what the actions are, whereas determinism means that there is only one course our actions can take in this particular world and of course they may have only one possible outcome, but it is not true that if we took other action, the outcome would still be the same.

On the other hand, if we think about fatalism as an idea that each person has predetermined fate that it is not able to change, it is consistent with what incompatibilists think about free will and what we can derive from their usage of the phrase "could have done otherwise." If "was able to do otherwise" entails "might have done otherwise," it means that (again, due to the law of contrapositive), in a deterministic world a person has never been able to do anything otherwise, because they have never might have done otherwise. Therefore each person has a predetermined fate that they are not able to change. The only difference between this idea of fatalism and the more common one presented in the previous paragraph is that usually the fate is taken as a supernatural concept, whereas here it is a consequence of determinism.

The argument presented here is called by Van Inwagen "the Consequence Argument." This is his initial formulation of it:

> If determinism is true, then our acts are the consequences of the laws of nature and events in the remote past. But it is not up to us what went on before we were born, and neither is it up to us what the laws of nature are. Therefore, the consequences of these things (including our present acts) are not up to us. (1983, 56)

In the book *An Essay on Free Will* he presents, as he says, three different versions of the same argument that is supposed to affirm the conclusion presented above, as he believes the initial formulation of the Consequence argument is "sketchy." Since these are three versions of the same argument, they are very dense and detailed and it is not necessary to consider all of them for the purpose of comparing incompatibilism and compatibilism, I will present here only one of them. The last argument presented by Van Inwagen (1983, 93–105) is especially relevant for the comparison in question, therefore let us have a look at it.

The third version of Van Inwagen's argument is called by him a "modal argument," because he employs in it a new modal operator that is quantified over sentences: N. Np means "p and no one has, or ever had, any choice about whether p." For example "N All men are mortal" means "All men are mortal and no one has, or ever had, any choice about whether all men are mortal." (Van Inwagen, 1983, 93). To make his conclusion Van Inwagen adds two rules of inference. Rule α is defined as

$$\Box p \to Np,$$

which can be translated into natural language as

"If it is necessary that p, then no one has, or ever had, choice whether p."

As Van Inwagen correctly notices, it is hard to reject this rule and not many philosophers, Descartes being one prominent example, are willing to ascribe a capability of choosing what is a necessary true even to God. The second rule, more controversial one, that Van Inwagen adds, is rule β:

$$N(p \rightarrow q), Np \rightarrow Nq.$$

It can be translated into natural language as

"If p entails q and no one has, or ever had, any choice whether p entails q, then if p and no one has, or ever had, any choice whether p, then no one has, or ever had, any choice about q."

And if no one has, or ever had, any choice about anything, we can hardly say anyone has ever had free will, no matter how we define it.

After introducing these two rules Van Inwagen makes an inference using them and assuming determinism that Np holds for any sentence p (1983, 94–95). I will not evoke the whole argument here – it is most important that the acceptance of the rule β is the most controversial part of this argument. One of the reasons why Van Inwagen thinks it is controversial is because depending on whether one accepts β one is either compatibilist or incompatibilist, but it can also work the other way around – depending on whether somebody is compatibilist or incompatibilist, one may or may not accept the rule β. The reasons for validity of β that Van Inwagen gives are his "intuition" and no counterexamples to β that he can think of (1983, 97–98). Actually the examples of β that he gives make him have the "intuition" that it is a valid rule of inference.

In the end β turned out to be an invalid rule of inference what was presented by McKay and Johnson (1996). They added a new rule of inference called Agglomeration that follows from α and β:

$$Np, Nq \rightarrow N(p \wedge q).$$

Since α is obviously correct, producing a counterexample to Agglomeration proves invalidity of β. McKay and Johnson prove invalidity of β in following way. Let us suppose that there is someone that did not toss a coin, but he could have done it. p represents the sentence "the coin does not fall heads" and q represents the sentence "the coin does not fall tails." Obviously, Np and Nq are true, as no one has a choice whether a coin falls heads or tails. But the conclusion N(p∧q) is false, because the person in question could have tossed a coin rendering p∧q false.

McKay and Johnson present their own, rectified version of β which employs a more concrete interpretation of what Np means in terms of ability. According to them, Np means

$$p \wedge Vxp,$$

where Vxp means

> "Every ability if x's is such that after every exercise of that ability, p."
> (1996, 119)

The new rule β has a form: p, Vp, p→q, V(p→q) → q∧Vq. The counterexample presented for the original β does not work for the new β, because in the original argument Np and Nq were true, whereas here Vp and Vq are false. The new β can serve well the purpose of the argument made by Van Inwagen and, unless somebody comes up with a counterexample to it, it is valid, which, according to McKay and Johnsons, entails that a compatibilist must reject "could have done otherwise" as a condition on free will or show that their usage of this notion is not adequate (1996, 121).

This argument (or arguments of this kind) is the most important argument in favour of incompatibilism. The whole debate between compatibilists and incompatibilists is based on an argument whether free will is compatible with the truth about the past and laws of nature determining everything what happens. There are two things worth to notice. First is that this argument does not support a thesis that free will is compatible with indeterminism – it may be incompatible with both determinism and indeterminism and there are arguments for that (discussed later). Second, this argument does not even need a strict definition of "free will" to work. Its aim is to show that determinism is incompatible with anybody having a choice about anything and if having a choice about anything is incompatible with determinism, then free will is incompatible with determinism, whatever it is. But it does assume there is a correct way of speaking about free will, in line with what Van Inwagen suggests about the free will problem: free will is either compatible or incompatible with determinism, there are no two distinct kinds of free will (libertarian and compatibilist) out of which one is compatible with determinism and the other is not (Van Inwagen, 2017, 154). All these things are employed by various critics of incompatibilism.

Free will and the consequence argument

Let us now discuss a few of the replies to the consequence argument in relation to our main topic – free will. After all, it is the most important what incompatibilism means for free will, human autonomy, etc., establishing that free will is incompatible with determinism (if successful) is only halfway through. In this section I will discuss arguments implying hard incompatibilism, as I will devote a separate section for arguments for compatibilism and, what is quite obvious, arguments for compatibilism are at the same time counterarguments to incompatibilism. But here I am concerned with only whether it changes anything for free will if incompatibilism is true. In other words, I will explore arguments against compatibility of free will and indeterminism.

The first important argument that is very often employed against compatibility of free will and indeterminism is that if indeterminism is true, then human actions are admittedly not determined, but this lack of determination means that there must exist randomness in the universe, but random actions cannot account for free will. Why? Because an agent that performs an action cannot be its cause or control it. If actions are effects of random evens (if it is possible to still call them "actions"), then it is just a matter of chance what action which agent will perform. This is an argument called by Van Inwagen "The Mind Argument," because it has often appeared on the pages of the Mind journal (Van Inwagen, 2017).

Let's take a closer look at the Mind argument, referring to Van Inwagen's remarks. What does it mean for an action to be a matter of a random chance? For that to be the case there must be a probabilistic law that describes behaviour of an agent. Let us say there is an agent a that has two choices: c_1 and c_2. A formal probabilistic description of this situation could take form like this: given circumstances a is in, there is $x\%$ chance that a will choose c_1 and $100\% - x\%$ chance that a will choose c_2, where x is a real number between 0 and 100. But, to make it more concrete, let us think about particular choices c_1 and c_2. Let us suppose that a is a person that wants to murder person b and c_1 is an event that a decides to pull the trigger of a's gun which results in killing b, whereas c_2 is an event that is the complement of c_1 – a decides to not pull the trigger and b is alive. Now, according to the way probability is understood by Van Inwagen (2017, 105–106), it means that if God made a to make the choice between c_1 and c_2 a thousand times (for example by reverting state of the world to the moment of choice), then a sometimes would choose to kill b and sometimes a would refrain from it. What Van Inwagen assumes, is that as the number of trials increases, the ratio between the number of times a killed b and the number of times a refrained from that should converge to some particular number (of course between 0 and 1), although he mentions a possibility of an improbable scenario, that the ratio does not converge. This argument against compatibility of free will and indeterminism is called a "Rollback Argument."

Now let us also assign probabilities to these two events – let us suppose that both c_1 and c_2 have probability of occurring 50% and after one thousand trials a chose about five hundred times c_1 and about five hundred times c_2. As Van Inwagen concludes, such an outcome should make an impression that in case of any subsequent trial the a's choice is just a matter of chance (2017, 106). It is surely not determined by the laws of nature and the initial state of the world, but it is also not an ability of a to choose what a will do. And what holds for all replays holds also for the first situation where a had to make a choice. The conclusion is simple: an undetermined action is a matter of chance.

The Mind Argument is the reason Van Inwagen calls free will a mystery. While he believes (or hopes) there is a fault in this argument, which he cannot spot, hard incompatibilists accept its results and claim free will is incompatible with both determinism and indeterminism and therefore it does not exist. It is probably the most important argument in the debate between compatibilists and incompatibilists, because it is supposed to show not only that if indeterminism is true, then our acts are random, but also that we do not have a clear notion of what free will is under the condition of indeterminism, which makes it hard to even speak about it. Incompatibilism first makes a point that whatever free will is, it is not compatible with determinism. But then it seems to turn out that whatever it is, it is also not compatible with indeterminism, therefore it is an empty notion.

Since this argument is so important, let us take a closer look at it and possible arguments against it. Franklin (2011, 201) characterizes the argument using two simple premises:

1. If an action is undetermined, then it is a matter of luck.
2. If an action is a matter of luck, then it is not free.

Of course, under the assumption of incompatibilism these two premises help us entail that no action is free. As Franklin further notices, the premise (1) does not make it explicit to what degree an undetermined action is a matter of luck, and the premise (2) does not specify whether an action being a matter of luck means that an agent could not exercise any control over it. The thing that probably bounds these two problems together is what it means that an action is a matter of luck – what is "luck"? Is luck always a matter of indeterminism? I believe it is worth to start with these considerations.

It is not clear what it is supposed to mean that an action is a matter of luck. On one hand it may mean that while an action A happened at a time t, at a point in the past it was possible that at t some other action, B, happened. But it would just be restating the antecedent in the premise – the idea of indeterminism. On the other hand we can explain luck by the relation to the agent of an action. Luck is then a situation when an agent did A, but he/she did not have a control over his doing A. But it is restating the consequent of the second premise. It is also possible that "luck" in both of the premises has different meaning. Whichever option we choose, we can reduce the whole argument to "if an action is undetermined, it is not free," but then the argument has a form of the conclusion. And a libertarian incompatibilist does not have to agree with it.

Franklin (2011) gives another explanation of why the Mind argument fails. One of them is placing indeterminism in a wrong time– after the choice or action, instead of at the time of the action. For example if somebody tries his/her best to overcome temptation, it is the trying that is undetermined, not whether it is successful. This counterargument is aimed at a version of the Mind argument presented for example by van Inwagen (1983, 142), where he makes us imagine a device with one button and two bulbs – red and green – which are indeterministically lighted after we press the button. We have a choice about pressing the button, but we don't have a choice about which bulb will be lighted afterwards. It is nevertheless not clear how the Franklin's idea could preserve agent's control, since we can always transfer the problematic randomness to the moment of decision or action. Then, whether a person tries hard or presses the button is a matter of luck and we are back to square one. Also, in van Inwagen's example we can associate the outcome of pressing the button with an event "either the red or the green bulb will be lighted," which is deterministic after we make a decision of pressing the button.

It is not hard to see that the main problem of libertarians is to correctly define the notion of having a choice that would work in an indeterministic scenario and this is probably one of the reasons for believing in hard incompatibilism. For purpose of completeness it is worth to mention here one allegedly sensible way to defend the notion of choice in an indeterministic world – the way that employs a so-called "agent causation." It is of course a way to say that an event may be not random and not determined at the same time – it is "agent caused." It is only an allegedly sensible way, because, as in case of making a decision when pressing a button, it shifts the problem of indeterminism one step earlier. This is what van Inwagen accuses defenders of agent causation of (2017, 110). To make it even more clear, we can say that agent causation is just a hypostasis. We want to answer a question "What does it mean that somebody had a control over a choice?" and an answer "It means that the choice was agent caused." is not instructive at all. We can even say that "an event y was agent-caused" is just an equivalent of saying "person x had a choice about y" for person x, but we still do not know how x could have a choice about anything if indeterminism is true. It may also be the case that actually the explanation of what agent causation is comes from explanation of free will. Denying free will may result in problems with a coherent definition of a person, so it is possible to imagine that we would rather say that an act was agent caused when it was free than the other way around.

Now let us talk about another response to the Rollback Argument. What happens if the probabilities of making each of the choices are different? Does it not change anything in comparison to the situation where both of the choices have probability 0.5? This is the path that is explored by Cogley (2015) in his article *Rolling Back the Luck problem for Libertarianism*. He considers three different scenarios when a person is in a hurry on the way to an important meeting that may have influence on his/her career but has to decide whether to intervene in an assault. In the first scenario the person, Anne, stops and intervenes and if God rolled back the situation multiple times, 50% of the time she would have chosen to stop and 50% to go. According to Cogley, Anne possesses as much freedom and control over what she does as it is possible if she is equally inclined toward either choice. He argues for it by introducing the second scenario: Jan, who is in the same situation, but she favours more stopping and helping the assault victim instead of going to the meeting. But since Jan is weak-willed, if God rolled back the situation, the observed pattern of choices would also show 50%/50% split. Here it is important to understand what it means to favour stopping over going to the meeting. Jan may have a desire to progress up the corporate ladder, but she repudiates it. She may have also decided in the past to help victims in similar situations. She may be worried of favouring her own interests more than helping others – so she feels guilty. Cogley claims that whatever the explanation is, Jan has significantly less control than Anne had, because she has more reasons to stop than to go and yet the ratio in the rollback scenarios is 50%/50%. Anne may be a person that has resolved to try to balance her career ambitions and the desire to help others and it is reflected in her ratio in rollback scenarios, while Jan would be upset to know she helps the assault victim only 50% of the time. The last agent Cogley considers is Stan. He is also strongly committed to help, but, unlike Jan, he helps the victim in about 98% of the cases. Since there is a margin of 2% cases when he goes to the meeting, Cogley concludes that he is in a position to do each of the actions. But he exercises more control over his choice than Anne and Jan. Also, assuming that it is right to stop and help the victim, Stan's decision to stop and help is more praiseworthy.

Does this really help in answering the Rollback Argument? Before answering this question there are two things to discuss. First, I am not entirely sure where the difference between Anne's and Jan's choices comes from. Cogley says that Anne decided to balance her career ambitions and helping people, while Jan favours helping more. But these resolutions can be viewed as just one among many reasons when they make the decision.

In the end it is important what they decide in the particular situation of choice, and it turns out that their inclination towards each of the possible decisions is the same. The argument could work better if we imagined that Jan decided to help in most cases, but suddenly changed her mind and walked away in many of them, which resulted in the 50%/50% ratio. But it would still beg the question whether she decided to help or not – what was the "final" decision? Another important thing are Stan's choices. Why does he need to sometimes act differently to be able to say that he is not determined to do what he does? Concretely, if in 1000/1000 scenarios Stan chose to help, does it mean that he is determined to help? I think Cogley (and not only him) mistakenly equates metaphysical problem of determinism and empirical question of assigning probabilities to decisions. If there is a possible world in which Stan does something different than he did, it still exists even if in rollback scenarios he makes only one decision. Whether there is such a world is a metaphysical question. How we will expect Stan to behave in future is a question that we can answer using empirical investigation, by checking how often Stan chose to help in similar situations in the past. We can say it is reasonable to expect he will do the same if we have to bet and in 100% of the scenarios in the past he did one thing, but it may not be unreasonable to say that we do not have enough information to exclude that at this particular time he will act differently. In other words, if determinism is true, he has to make the same decision in each rollback. But if he makes the same decision in each rollback, it does not mean that determinism is true or that his choice is determined. Cogley even explicitly denies it by saying:

> [O]n the account I am developing, a libertarian must require an indeterministic link between a person's mental set and her choice. So the libertarian must balk at attributing full control when rollbacks show a person doing the same action 100% of the time. (2015, 135)

He also says that somebody objecting his argument might say that there is nothing libertarian in it, because a person perfectly inclined to do something would always choose the same option and it would *mean* he/she is determined to choose it. I believe the distinction between being determined and performing the same action in each rollback is a subtle, yet particularly important, and it is very often overlooked by people engaged in the problem of free will. I would say it stresses what is in the core of the problem and I will talk more about it in the last chapter of this thesis.

Given the two things I discussed in the previous passage I do not think everything is clear in Cogley's argument, but he may be right we need to look at different ratios of decisions when investigating whether the Rollback Argument poses a threat to libertarians. Yet, it is still not clear what it means to have a control over an action – we still need a positive argument that would help in establishing it or maybe it is enough to argue that a person has a control of an action when he/she is not determined to choose A over B, but yet most of the cases he/she choses one of the options over another, just as Stan. Cogley argues that the rollback scenarios are not only harmless for libertarians, but they can show the degree of control of an agent over an action. This is an interesting path to follow, but it would definitely need clarification especially because of the problems with conflating determinism and probability of an event being equal 1.

The last argument against the Rollback Argument that I would like to discuss refers exactly to the problem of interpreting probability. Lara Buchak, in her article *Free Acts and Chance: Why the Rollback Argument Fails*, instead of arguing why the fact that some act has some concrete probability of being chosen by some person does not diminish his/her control over the action, she argues that we do not have any reason to assume that in the rollback scenarios probability would converge to some particular number between 0 and 1 – it can diverge. Van Inwagen in his original statement of the Rollback Argument says that the ratio of choices will "almost certainly" converge. As he says "[I]t is possible that the ratio does not converge. Possible but most unlikely: as the number of replays increases, the probability of "no convergence" tends to 0" (2017, 106). As Buchak correctly notices, there is no reason to think so – actually van Inwagen does not even point us in a direction of any possible explanation. She suspects that the reason for "almost certainly" is because of mistaken usage of the law of large numbers (2013, 23). Here is what she gives as a typical statement of the law:

> In repeated, independent trials with the same probability p of success in each trial, the percentage of successes is increasingly likely to be close to the chance of success as the number of trials increases. More precisely, the chance that the percentage of successes differs from the probability p by more than a fixed positive

amount, e > 0, converges to zero as the number of trials n goes to infinity, for every number e > 0. (2013, 23–24)

For example if I flip a fair coin many times and I want to see the ratio of flipped heads to flipped tails to be between 0.45 and 0.55, when the number of my trials goes to infinity, I am guaranteed (the chance that it is outside this interval converges to 0) that at some point it must be between these two numbers. But in case of flipping a fair coin we know what the probability of getting tails or heads is – it is 0.5. If a coin is unfair the probability for each of these events to happen is different and we may use the law of large numbers to determine what the probability is. But in case of flipping a coin we assume there is *some* probability for each of the events. If van Inwagen assumes there is such concrete probability for different choices in the Rollback Scenarios, he assumes something he wants to prove – that if indeterminism holds, then when a person has to decide between A and B, there is an objective probability for him/her to choose A and an objective probability for him/her to choose B.

Schlosser (2016) tried to show that even if Buchak is right about van Inwangen's missing reason for assigning probability to the choices, there are other reasons that may make us do it. He says that personal reasons for choosing each of the actions must reflect the probabilities of each choice. Otherwise, he asks:

> How else could one construe the influence of reasons over undetermined choices and actions in a way that can reflect the varying degrees in their normative strength? (2016, 342)

I will not try to answer this question at the moment, but I believe that addressing this issue is important for any libertarian account of free will. As we could see here there is one main problem with various arguments for incompatibilism. They do not help us determine what free will actually is – they only let us say why it cannot exist in a deterministic universe – and Schlosser's reply to Buchak is probably aimed at providing a positive libertarian account of free will. The discussion over the Rollback Argument also shows that both proponents and opponents of indeterminism or libertarianism have problem with a basic notion that has to be involved in discussions of these two terms: the notion of probability.

Compatibilism

Compatibilism is a stance which states, of course, that free will is compatible with determinism. Before digging deeper into one of the contemporary views on compatibilism, let us take a look at a historical case. One of the first proponents of compatibilism was Hume, although he did not call himself this way. In his *Treatise of Human Nature*, Hume writes about the will in general to be nothing more than "the internal impression we feel and are conscious of, when we knowingly give rise to any new motion of our body, or new perception of our mind" (Hume, 2009, s. 612). His idea of freedom of will – or "liberty", as he calls it – is best described in his other book, *An Enquiry Concerning Human Understanding*: "By liberty (…) we can only mean a power of acting or not acting, according to the determinations of the will; that is, if we chuse to remain at rest, we may; if we chuse to move, we also may" (Hume, 2007, 69). His basis for this stance is a claim, that it is obvious, that human actions are based on motives, inclinations and circumstances. They are causes for human behaviour and, since the idea of cause assumes necessity, they necessitate actions, while being necessary themselves. Free will described in this way can be thought of as an idea composed of two simpler ideas: freedom and will.

Contemporary compatibilists did not go much beyond the core of this Humean idea of free will. Here we can already see the first advantage of compatibilism – it very quickly and easily provides us with a notion of free will that is reconcilable with determinism. As it was mentioned before, van Inwagen believes that compatibilists are people who believe that from the fact that somebody could have done otherwise we cannot infer that that person *might* have done otherwise at the same time. It should be quite clear that compatibilists believe that this entailment does not hold, because determinism implies that at any given state of the world, there is only one possible world accessible from it. So if determinism is true, then definitely nobody might have done something different than he/she did. What matters whether a person was able to do otherwise (it was within his/her power to do otherwise, etc.) and the fact that he/she was determined to actually not do it does not matter.

So now it is at least clear what it means to have free will when doing something. I have free will when I am uncoerced to do what I do. It is simple, but the question is how convincing it is. The whole previous subchapter was devoted to arguments for incompatibilism, which are (sometimes directly, sometimes indirectly) also arguments against this notion of free will. Let us now trace the arguments for compatibilism, which are *de facto* arguments for a specific understanding of what free will is.

Arguments for compatibilism

Let us start with considering why it does not matter if a person might have done otherwise or not for him/her having free will. This idea is explored by Dennett in his article *I Could not have Done Otherwise – So what?* (1984). Although the title refers to the ambiguous "could have done" term, it is clear that Dennett means "might have done" here. He says, "I assert that it simply does not matter at all to moral responsibility whether the agent in question could have done otherwise in the circumstances." which makes it quite obvious that he talks here about an agent being in *exactly* the same circumstances.

First Dennett's argument is that when a person does something that he/she can be blamed for, we never investigate it thoroughly whether he/she could have done otherwise in that situation. As Dennett admits, it may be a startling statement, but the word "thoroughly" is the key here. As Dennett, quite rightly, notices, ordinary people never withhold judgments about responsibility until after they have consulted a physicists or other scientists that could tell them about determinism possibly being involved. What they check is whether they can exclude so-called "local fatalism" (1984, 555). Local fatalism is a term coined by Dennett that refers to a situation in which no matter what somebody does, wants or tries to do, the effect is the same (as in case of a person locked in a room). As Dennett says, this kind of fatalism is compatible both with determinism and indeterminism. Local fatalism can be an explanation of a person's behaviour that makes it impossible to hold him/her for a deed. But when ordinary people try to ascribe responsibility for acts, they do not go beyond local fatalism and when it is excluded, a doer can be held responsible.

The second point that Dennett makes is that sometimes we draw the conclusion opposite to what indeterminists conclude when somebody could not have done otherwise. Namely, the fact that it was impossible for somebody to do what he/she did is a reason to praise him/her. He gives here an example of Martin Luther who said "Here I stand. God help me. I can do no other." when concluding his testimony before the Diet of Worms." For Dennett what Luther did shows that we do not exempt someone from blame or praise just because he/she could do no other. He confronts this situation with another situation when somebody admits he/she could do no other. If a person who has an irrational fear of flying refuses to get on a plane that would transport him/her to safety and says he/she can do no other, for Dennett it is a sign of an impaired control and an excusing condition. But it is not so in Luther's case – it is his rational control faculty that makes it impossible to him to act differently. The reason dictated what he had to do, and he would have to be mad in order to not do it (Dennett, 1984, 556). Now, Dennett concludes, if "could have done otherwise" was needed for moral responsibility, on each occasion when somebody makes a decision out of reason and that decision is for him/her unquestionable, we would have to diminish that person's responsibility for the decision.

The last argument that Dennett makes refers to possibility of determining whether an act was determined or not (1984, 558). He claims that due to the brain's complexity it is practically intractable and if so, it is a big problem of ignorance – if "could have done otherwise" is, again, a condition for moral responsibility, we can never know if somebody is morally responsible for anything.

All these arguments are used by Dennett to conclude that using "could have done otherwise" principle to determine responsibility is misguided. He claims that not only we cannot know if anybody ever could have done otherwise, even if we could know it, it would not give us any important knowledge. What we really need is to determine whether an agent could have done otherwise in *similar*, not the same, circumstances. It is even more important when we realize that it is impossible for a human being to be twice in exactly the same situation – even if environment were the same (which is also impossible in principle), the state of the human being would be different because of the previous experience. Dennett compares here an agent to a robot designed to live its entire life as a deterministic machine on a deterministic planet (1984, 559). When the robot makes a decision, it is an outcome of built-in heuristics that trigger pseudo-random selection of behaviour at various points. Whatever it chooses, it could not have done otherwise by definition – when we talk about metaphysical sense of "could have done otherwise." But, as Dennett says, there is another sense in which we can talk about it. If the robot makes a regrettable mistake, its designers may ask "Could it have done otherwise?" and think if they could have designed the robot in a different way, which could have prevented the robot from making the mistake. But what they are interested about is to make sure that the robot will not make a similar mistake in the future. If robot's heuristics and other features programmed into it turned out to be correct, they might say that the robot could have done differently in a similar situation – its behaviour was caused by circumstances that could not have been predicted.

The result of these arguments is that, according to Dennett, we ask the question "could he/she have done otherwise?" because we want to interpret something that has happened to draw conclusions about the future. We want to know whether we can trust an agent that he will or will not behave the same way in the future again, when the circumstances are similar, but of course not identical.

It is supposed not to change anything in our ascription of praise and blame. Dennett claims that there is no reason not to blame somebody for an act he/she could not refrain from doing, because being determined to do something does not make any difference. He considers an accusation that someone might make about it – if a person was determined to do something, he/she had no chance not to do it. For Dennett it works only for an extremely superstitious view of what a chance is (1984, 564). To make his point he compares two lotteries. In Lottery A the winning ticket is drawn after all the tickets are sold, in Lottery B before. There are people that think that Lottery B is unfair, but actually the time of selection does not matter for fairness of a lottery – in the end the chance of winning is for everyone the same. So, the concept of chance itself is valid both in case of determinism and indeterminism.

These are the main thoughts presented in Dennett's *I Could not have Done Otherwise – So What?*. I believe this article serves well as a basis for discussion about compatibilism, as it shows well the shift in thinking about free will present in deliberations of compatibilists in comparison to what we have seen in case of incompatibilists. Even though Dennett wrote multiple books on the topic or mentioning the topic afterwards, his position as a compatibilist did not change much and everything is based on the same convictions as I presented above. The difference that we can definitely spot is in the language and the idea of free will itself is clarified, as here we could only see why the fact that somebody could not have done otherwise is supposed to not change anything for free will and his/her responsibility.

There is one important extension to what was said above, and we can find in the book *Freedom Evolves* (2003). Dennett employs there the theory of possible worlds in a way that makes it possible to him to say that somebody could have done otherwise even if determinism is true. As it was already stated, if determinism is true, there is only one possible future accessible at any time of existence of this world. In terms of possible worlds there exists only one possible world accessible from this world at each instance of existence of this world. For Dennett it does not change that when a person P chooses option A instead of B, we can still say that P could have chosen otherwise. Why? Because although in our world P chooses A, there are worlds that are very similar to our actual world and in which P chooses B (or maybe P may choose either A or B, because these worlds could be indeterministic). It is supposed to work, because according to Dennett we think in terms of possible worlds not because our world is indeterministic, but because we do not have complete knowledge of our universe. As he says (2003, 69), supposing that determinism is true and we know all the laws of physics like the Laplace's demon, if we do not have perfect knowledge of the state of the universe at any time, we would still not be able to say which world will actualize as the next instance of this world. Without this knowledge we can engage ourselves in talking about possible worlds even if determinism is true.

As an example of application of possible worlds Dennett presents talking about counterfactual sentences. One of them that is discussed by him is "If you had tripped Arthur, he would have fallen" (2003, 69). According to Dennett, this sentence is true if in every world that is approximately similar to our own, if the antecedent holds, then also the consequent holds. As he adds, sometimes when we make such counterfactual claims we think about different possibilities, which could prevent the consequent of being true even if antecedent is true. In case of tripping Arthur we could imagine that he would not have fallen, even after being tripped, if "the room was filled with inflated air bags or the whole building was in free fall with zero gravity," (2003, 69) but Dennett claims this world is too dissimilar to count.

The last argument for compatibilism that I would like to briefly mention is a conclusion of the Mind argument made in favour of compatibilism. Here is how van Inwagen sketches the Mind Argument:

> If one's acts were undetermined, they would be "bolts from the blue"; they would no more be free acts than they would if they had been caused by the manipulations of one's nervous system by a freakish demon. Therefore, free action is not merely compatible with determinism; it entails determinism. (van Inwagen, 2017, 40)

The conclusion here is of course opposite to the conclusions made by incompatibilists – if any act is free, then it is determined. This is important to stress – this argument does not help in establishing that free will exists even if determinism is true, which is a necessary condition of free will according to this argument. It is similar in structure to what incompatibilists had to say about free will, but the compatibilist's advantage is a clear notion of free will.

Discussion of arguments for compatibilism

It is easy to notice that the arguments made by Dennett have quite a different form than the arguments for incompatibilism described before. While the arguments for incompatibilism have logical form and are carefully prepared formally, the arguments for compatibilism are grounded rather in a sociological or even quasi commonsense analysis. Their aim is not to show that a coherent discussion involving free will presupposes compatibilism (as it is in case of incompatibilism), but rather that it is impractical to claim that free will presupposes indeterminism. On the other hand the Mind argument has a more formal character, and it is possible to tackle it this way, so let us start with showing a reply to it before getting deeper into discussion of sociological issues that Dennett evokes in favour of compatibilism.

In an essay *On Two Arguments for Compatibilism* van Inwagen makes a claim that the Mind Argument is incoherent with the so-called Ethics argument (2017, 40–42). The Ethics just states what "could have done X" means for a compatibilist: "the correct analysis of 'X could have done A'

is 'If X had decided (chosen, willed...) to do A, X would have done A'."
(van Inwagen, 2017, 40). Van Inwagen claims that if this is sound, then
the Mind argument is unsound and the other way around. His argument is
based on the fact that the Ethics Argument makes free will compatible not
only with determinism, but also with indeterminism. Let us suppose there
is a person X whose choice of lying or telling the truth is undetermined,
X chooses to tell the truth, but if X had decided to lie, X would have lied.
Probably different decisions could be random events or "bolts from the
blue," but it does not change anything according to the Ethics Argument,
as it does not specify what the source of what a person wants is. But if it
is true that if X had decided to lie, X would have lied, then X acted freely
and free will is compatible with indeterminism. But, according to the Mind
argument, free will presupposes determinism. So, Van Inwagen concludes,
the arguments are incoherent.

One way to answer this counterargument would be to say that whereas
the Ethics Argument makes an initial condition for free will, which is compatible both with determinism and indeterminism, the Ethics argument
gives the further condition by ruling out possibility of free will if indeterminism is true. If so, there is no incoherency between the two arguments
and van Inwagen is not right. But on the other hand if according to the
Mind Argument free will is impossible if indeterminism is true, then the
definition of free will compatible with this argument must be different than
the one used in the Ethics Argument, even though the second one was not
specified. It would then be possible to formulate the Mind Argument in
a way supporting incompatibilism: "If one's acts were determined, they
would be a result of some events in a distant past (before one was born)
and the laws of nature; they would no more be free acts than they would
if they had been caused by a pseudo-random generator of numbers built
into the brain of a person at the time of birth. Therefore, free action is not
merely compatible with indeterminism; it entails indeterminism." But this
argument is unsound. From the fact that free will is incompatible with determinism we cannot derive a conclusion that it entails determinism, not
even that it is compatible with indeterminism, and it is of course the same
for trying to derive that free will entails determinism if it is incompatible
with indeterminism. In both cases we do not see a coherent notion of free

will that could be supported by either determinism or indeterminism and we are back to square one.

Let us now get back to the arguments evoked by Dennett. Two of them are supposed to help to establish that it does not change anything in responsibility for a deed of a person if a person could not have done otherwise – either because people do not investigate thoroughly whether a person could have done otherwise when assessing responsibility or because it is not even possible to do. Both of these arguments do not sound to be philosophically sound. If ordinary people do not investigate thoroughly whether a person could or could not have done otherwise, it only gives us information about them and their habits – not about any link between free will or "could have done otherwise" condition and responsibility, especially when each of these terms is highly ambiguous. It should not be a surprise that "ordinary people" make decisions that are not based on exhaustive philosophical arguments or use ideas that are contradictory and do not even notice it. It is a task for a philosopher to show that what "ordinary people" do or think makes no sense, if it is really so. From the fact that ordinary people, while ascribing responsibility, do not thoroughly check whether a person could have done otherwise we cannot possibly infer that it does not matter for moral responsibility. There are therefore two possibilities here. Either Dennett makes a terrible philosophical mistake, or it is a hint that his aims are not philosophical. I will argue later that we have to assume the second possibility. It is similar in case of the argument that even if we cared if someone could have done otherwise, it is impossible to know it for sure. Namely, if Dennett is right and we cannot know it, maybe something is wrong with our notion of responsibility or with our ascriptions of responsibility. Van Inwagen, however, discusses this issue in more detail in his *Critical Study of Dennett's "Elbow Room."* He makes us imagine "rogs" – robotic dogs that we cannot in practice distinguish from standard dogs. Then, he introduces a principle DR: "A thing is a dog only if it is not a rog" (2017, 55). The argument reads:

> It is in practice impossible to find out whether a (superficially doggish) thing is a rog or a dog. If DR were correct, therefore, we could never know whether a (doggish) thing was a dog ... Therefore, if there is any concept that goes by the name "dog" and which satisfies the demands of DR, it is not the concept that goes by that name in everyday life. (2017, 55)

The problem with this argument is that it makes a premise that if p entails q, then we cannot know that p unless we get to know that q and the same problem is shared with the argument Dennett makes. We are not in practice capable of distinguishing rogs and dogs, but it does not change anything in our concept of a dog and the same applies to moral responsibility and free will.

Another Dennett's argument refers to the fact that we do not exempt people from moral responsibility when they claim they could not have done otherwise, because it would be unreasonable. I think it is instructive to discuss this argument more thoroughly, as it shows, I believe, deep misunderstanding between compatibilists and incompatibilist. Dennett may be right that we do not exempt people like Luther from moral responsibility, but he seems to be completely wrong about reasons for which we do that. Here the confusion between "was able to do otherwise" and "might have done otherwise" as possible translations of "could have done otherwise" is very visible. I believe we do not exempt people like Luther from moral responsibility because what he means by "I can do no other" means "If I wanted, I could do something else, but I would never want it." He was physically capable of doing something else. What is important is whether he had access to a possible world in which he would do something else, even if he did not want to use that option under any circumstances. But if Luther was literally unable to do otherwise, then maybe, again, our ascription of praise and blame are wrong and we should say "he had to do this, nothing to talk about, we should not praise him for this"? The idea that Luther was unable to do otherwise is actually incoherent with compatibilism itself – if he wanted, he could do otherwise. For Dennett, the fact that Luther would never do something different in the same circumstances means he could not have done otherwise. For some incompatibilists this also holds – if a person chose one option in all rollbacks, it would mean he/she was determined to do it. But it does not have to be the case. There still may be some physically possible world in which that person does something different, he/she has access to it, but never uses it. In this way compatibilism and incompatibilism are compatible. If a person wanted to do otherwise in this world, he/she would do it, it is physically possible that he/she does it, but it never happens.

The next thing is the change in how Dennett thinks about possible worlds as an argument for compatibilism. According to his interpretation, a possible world does not have to be accessible from the actual world to be taken into consideration when we consider whether someone could have done otherwise. It is enough that it is accessible from a similar to the actual worlds. Of course the main problem here is to define the notion of similarity. But the problem is that even if we, in our generosity, assume that it is possible to define it and use in a consistent way, this idea can be easily used to support incompatibilism. Let us imagine a meteor hitting the Earth and destroying one of the Earth's cities. Someone may say, while looking at the big hole in the ground, "if the meteor did not hit the city, the city would still be here." Was it possible that the meteor did not hit the city? If the world is determined, there was no possible world in which the meteor did not hit the city. Yet it is of course nothing strange to say that if it did not hit the city, the city would still exist. We can explore possibilities to say what could prevent from extinction of the city. One simple idea is that there could have been another meteor that hit the first one and changed its trajectory. Since this is my example I can add here whatever I want, but it is not the point – the point is that the change of trajectory can never have its source in the meteor itself. So we cannot say that the meteor was able to not hit the Earth. But we can say the meteor *might* have not hit the Earth. There is a possible world similar to the actual one, in which the meteor did not hit the Earth. Why does it matter? The meteor does not have free will, that is sure. One can say that there is a big difference between meteors and people in the way how their "actions" are formed. But this would be to assume something that one wants to prove. Dennett wants to prove that our understanding of agency does not change whether determinism is true or not. But the problem that incompatibilists try to bring to the light is, in the end, that if determinism is true, then what we call "actions" are actually "events." We can easily say there is no difference between the meteor not hitting the ground and a person not performing his/her act, because to make these things possible, environmental changes are needed. In case of the meteor something needs to change its trajectory, in case of a human being something needs to influence his/her brain to function differently so that the decision made is different. And counterfactual conditionals do not change here anything.

Finally, Dennett makes a point on why we actually ascribe responsibility for deeds. We want to know whether an agent could do something otherwise in similar circumstances, not the same ones (as the same circumstances are never possible), because we want to prevent him/her from doing the same thing in the future, if it is possible. It makes it clear that when we, as a society, punish somebody for what he/she did, we do not do it as a retribution for what a wrongdoer did any more[3]. We want to change the future, not establish in detail what happened in the past – might he/she have done otherwise or not? It is hard to overstate the importance of this shift. Dennett is not interested in a metaphysical issue – how human agency and determinism are related to each other – but in a practical one – what we can do to make people behave in the "correct" way. He makes it clear that determining whether somebody was determined or not when doing does not change anything (1984, 564).

All these things are compatible with each other, but it is hard not to have an impression that they are not compatible with what arguments for incompatibilism refer to. Van Inwagen's arguments are supposed to answer metaphysical questions, while Dennett tries to show social consequences of speaking about free will in a particular way. His aims are especially clear in the books *Elbow Room: The Varieties of Free Will Worth Wanting* (1984) and *Freedom Evolves* (2003). In case of the first one even the title gives us a clue about it – Dennett assumes there is no one "free will," there are many and some of them, particularly the one he endorses, are worth wanting. The one incompatibilists endorse is not something we should care about, even if it exists. In *Freedom Evolves* he compares "genuine" free will to unhealthy "real" butter that can be replaced by margarine:

> What makes the "genuine" varieties worth caring about at all? I agree that margarine isn't real butter, no matter how good it tastes, but if you insist on real butter at any price, you really ought to have a good reason. (2003, 225)

---

3   E.g. "Yes, we should shed the cruel trappings of retributivism, which holds people absolutely responsible (in the eyes of God) for their deeds; we should secure in its place a sane, practical, defensible system of morality and justice that still punishes when punishment is called for, but with a profoundly different framing or attitude." (Dennett, 2017, 381)

Why he insists on using "artificial" free will is clear in his speech towards the scientists, where he argues why scientists should not tell other people that free will does not exist[4]. Scientists should not tell other people they do not have free will, no matter whether it is true or not, because it will change people's behaviour. Dennett seems to be here the least concerned with philosophical and scientific accuracy and more about psychological and sociological consequences of making some views available to the general public. If we consider his work from this point of view, it starts making much more sense and I believe this is the way how we should look at Dennett's theses.

Summary

When discussing compatibilism and incompatibilism, I decided to mainly focus on two prominent examples of compatibilism and incompatibilism – advocated by Dennett and van Inwagen, respectively. I decided to include a short discussion of these stances in this thesis, because I believe a thesis about human agency without referring to problems highlighted by compatibilists and incompatibilists would be incomplete. But there is no reason to get deeper into this topic. I believe there is no difference in opinions about free will between compatibilists and incompatibilists – they talk about completely different things.

In case of compatibilists' beliefs there is even no question whether free will is compatible with determinism. It is, because its definition makes no room for any incompatibility. Van Inwagen believes it is the same free will that compatibilists and incompatibilists talk about. He opposes talking of "incompatibilist free will" and "compatibilist free will." As he says, if somebody was an incompatibilists and then becomes a compatibilist, he/she would not say "I believed free will was incompatible with determinism, but I was wrong, free will is something different than I thought and this thing is compatible with determinism," but rather "I believed free will was incompatible with determinism, but I was wrong, the very same thing I called free will is compatible with determinism." (van Inwagen, 2017, 156). On the other hand Dennett explicitly states there are multiple

---

4 https://www.youtube.com/watch?v=vBrSdlOhIx4

varieties of free will we may want, one that incompatibilists want does not exist or is impossible and one what Dennett wants exists and is a completely good free will to believe in (Dennett, 2003, 225). Actually Dennett goes even further – it does not matter whether "incompatibilist free will" exists or not, the question is what *pragmatic* reasons we have to believe in it.

The postulate of van Inwagen makes sense and I believe that following it would be the only way to make the current dispute between compatibilists and incompatibilists sensible. I believe it is impossible and here I will give a few arguments why I believe the dispute is futile and make a clear statement how I want to talk about free will in the subsequent chapters of this thesis.

First of all – language is not determined enough to account for the subtleties that could help us decide whether free will is compatible or incompatible with determinism. In the discussion of arguments for incompatibilism we can clearly see that proponents and opponents of incompatibilism try to show different scenarios in which we could say that somebody "might have done something," "could have done something," "was able to do something," and from this derive a conclusion that free will is compatible/incompatible with determinism. But it is perfectly possible that these cases are not entirely clear and sometimes a philosopher must make a decision about definitions. Even our idiolects are different, should it lead, in extreme cases, to different views on free will? When we talk to other people and in our conversation it turns out that we attach different meaning to the same words, debating over the real meaning of a word would be impossible in many cases. It may be different in case of natural kinds, like water, but in case of abstract entities like free will it seems to be obvious that our definitions may differ, and they always will have to be specified more concretely than what personal usage allows for. Also, if we stick to natural languages when we decide whether free will is compatible or incompatible with determinism, then it may turn out that whether human beings have free will or not depends on the language they use.

Language analysis can also lead to very peculiar results. We can say a fallen tree was responsible for an accident. Is it a figure of speech or do we really hold trees responsible? Maybe it is a kind of responsibility that is not worth wanting? When we make voluntary actions, we move skeletal muscles. It is common to think that if somebody was threatened, he/she is not fully responsible for his/her action. Let us imagine that a person A steals something valuable because of a threat that if A does not do this A's child will be killed. In order to steal, A needed to move A's skeletal muscles, therefore the action was voluntary, from some certain point of view, we can say. Does it not mean that A is responsible for it even though A was threatened? We can probably give hundreds of similar examples based on linguistic analysis and it would not change much in our understanding of free will and responsibility.

Second, and more important, – even if it was possible to use linguistic arguments and analysis to make it clear that what we call free will is compatible or incompatible with determinism, what does it matter? Compatibilists or incompatibilists could say "yes, we were right, he/she could/could not have done otherwise," or "yes, we were right, free will is compatible/incompatible with determinism," but what is important are the philosophical consequences of both possibilities. Consequences of compatibilism and incompatibilism may be entirely different. Dennett wants free will that will make it possible to judge people for what they do to do some kind of social engineering using punishment. If, on the other hand, incompatibilism is true, we may want to derive from it responsibility that allows for retributive justice. In general, different ways of thinking about the role of human beings in the world and the nature of their interactions with their environment will lead to a completely different view on their responsibility.

I believe this is the ultimate goal when talking about free will. It does not matter whether we can say "yes we have free will," when we do not know what change it makes for human agency and responsibility. To say it even more strictly, the aim about introducing the notion of free will is to stress the difference between actions performed by an agent and events an agent is a part of. Every idea of free will should be assessed based on whether this distinction is possible or not. But to achieve this goal it is necessary to start from a different perspective than compatibilists and incompatibilists do. It is probably obvious that the answer to the mind-body problem strongly influences what we think about free will. Even the Dennett's compatibilism stems from the idea of materialistic solution to the psychophysical problem, as I will later show. Therefore, in the next three chapters I am going to discuss what human agency is (if it is even possible to be preserved) when we assume some of the solutions to the psychophysical problem. Chapter 2 will be devoted to dualism, Chapter 3 to materialism and Chapter 4 to transcendental idealism.

Chapter 2: Free will and dualism

Introduction

In the first chapter I presented an overview of the contemporary conceptual framework used to talk about "free will" and in the end argued that it is too high level to give an exhaustive solution to the problem of human agency. Because it is underspecified, the solutions to the problem are only seemingly related – in reality they refer to completely different ideas. Therefore I suggested that we should change the approach to the problem. Instead of trying to answer general questions like "what is free will?" or "are humans free?," etc., we can gain much more from starting with a solution to the mind-body problem, even as an assumption, and then trying to make sense what human autonomy or freedom could be. Of course stating the problem this way would also be very unclear and would allow abuses. To make it stricter, my main interest will be criteria of how we can distinguish actions from mere events.

I assume it is undeniable that everything that happens in the world is an event. But it is not always clear (not in case of every metaphysical view) that some of the events are actions. Someone could say that this distinction is not clear either and I should explain what exactly events and actions are. But I do not think it is necessary at that point. The idea is that assuming various metaphysical conclusions regarding psychophysical problem will allow to define events and actions in a different way. But the most important here is not what definitions we make, but what the way of distinguishing actions from events is. For example, if someone says human beings have moral responsibility, I want to know how it is different than saying that a tree was responsible for an accident. Even though it may seem trivial, I believe there are stances that do not allow for such a distinction. I understand this distinction for actions and events is not clear at the moment, but it is impossible to make it more concrete without assuming some particular view on the nature of the relation between mind and body.

I am going to start with discussing human agency with regard to dualism, because it is where it has all begun for contemporary philosophy of mind. As we will see, many contemporary philosophers define their stances in relation to dualism, especially Cartesian dualism, if not by endorsing it (which is actually rare), then by opposing it.

Cartesian dualism

The most famous version of the so-called mind-body problem was probably stated by Descartes, in his *Meditations On First Philosophy*. This work is important for several reasons. First of all, it changed the orientation of philosophical investigations. Descartes emphasized all the problems that we encounter when we talk about the reality of objects that are outside us – the outer world in general. He realized that what we can talk about are states of ourselves and we have no right to believe that the reality is just as we see it. For Descartes, the only reason to believe that our senses do not deceive us is to believe in God who, as a good Creator, must have equipped our bodies with tools that tell us the truth about the reality. His observation that we have access only to states of our minds, not to the world itself, started a new era in philosophy with the main problem of the relation between the mind and the outer world and all other problems that can be derived from it, like problem of solipsism, problem of mental states of other people that are perceived in experience or the relation between mental states and the body that these mental states at least seem to control (since the body is also part of the world outside the mental states, it has to be distinguished from the problem of the relation between mental states and other parts of the outer world), to name a few of them. The problems that Descartes' talked about define the area of philosophy of mind also today. It is hard to imagine a convincing theory of mind that would omit a number of the issues or areas mentioned in *Meditations*. Even if there are philosophers that do not agree with Descartes nowadays (probably the majority of the contemporary philosophers), many of them[5] define their views by contrasting them with Cartesian Dualism and showing them as an opposition to this stance. Of course the most obvious philosophical reason for that are various problems that Cartesian Dualism encounters, but there are also ideological reasons – Cartesian Dualism is

---

5   E.g. Daniel Dennett (1991) or John Searle (2004).

not naturalistic. With this in mind it is time to outline what Cartesian Dualism actually is.

Cartesian Dualism is a stance in a psychophysical (mind-body) problem. The psychophysical problem in its simplest form is a question about how mind and body are related to each other. According to Descartes there are two completely different substances: extension and mind (*res extensa* and *res cogitans*). The essential property of extension is, of course, that it is spatially extended, which means that it has length, breadth and depth. As Robinson (2016) correctly points out, it is a big difference when compared to Aristotelian tradition. According to Aristotle, properties of matter are accounted for by form that it takes[6]. In Cartesian Dualism it has properties on its own which can be investigated by exact sciences. On the other hand, the essential property of mind is thinking (which is understood very broadly, for example it encapsulates also perceptions, sensations and feelings). Then, since the mind does not have material properties and extension does not have mental properties (in other words they do not share anything in common), then they cannot be the same substance and that is why this stance is also called a Substance Dualism.

Actually, we can say that Cartesian Dualism is not only a stance in psychophysical problem, but also what caused it in the first place. Although philosophically interesting, a question "what is the relation between my chair and the matter it is built of?" is already answered in the question itself. We can talk about the relation between the idea of a chair and its exemplifications in the material world, but it does not change much – still, my chair is a purely material being and its relation to the matter is quite simple, the abstract idea of a chair being a completely different thing. When we ask a question "What is the relation between mind and brain?" it is not surprising that the very first idea that comes to mind is that they are two different entities. The problem with Cartesian dualism seems to be not with arriving at the solution, but with creating a coherent picture

---

6 Of course this statement raises a question about the relation between Aristotelian matter and Cartesian extension, which falls outside of the scope of these considerations. It is nevertheless worth to point out here, that what Descartes calls *res extensa* could be an equivalent of a combination of form and matter in Aristotelian ontology.

of the relation between mind and body afterwards. But before digging deeper into it, let us examine the arguments that Descartes had to support his stance.

Arguments for Cartesian dualism

Arguments for Cartesian dualism must pertain and support the idea that there are two different, autonomous substances – material and immaterial, to speak about them very generally for now – and everything that exist must be an instance of one of them. In *Meditations on First Philosophy*, in the sixth meditation Descartes says that everything that can be clearly and distinctly understood is capable of being created by God exactly as it is understood (2008, 55). Therefore, if one thing can be clearly and distinctly understood without another, then it is a sufficient argument that they are distinct, since they can be separated at least by God. It is so in the case of mind and body, since the essence of mind is thinking and being a non-extended thing, and the essence of body is that it is an extended, non-thinking thing (which is actually a negation of being a mind). What applies to the difference between mind and body applies also to the difference between the mind and the faculties of thinking – imagination and sensation. Descartes says he could understand them without the thinking substance in which they inhere, but to understand them without the thinking substance is impossible, therefore the relation between the mind and these faculties is like between a thing and modes in which it can be. I believe it is not a coincidence that Descartes mentioned imagination and sensation as the faculties without which he could still clearly and distinctly conceive of himself – these are faculties whose working strongly depends on the interconnection between mind and body. But if the mind were disconnected from body, imagination or sensation would not be possible, but it would still be the same thinking substance.

Another argument that supports the idea that mind and body are different substances is an argument based on Leibniz's principle of indiscernibility of identicals, according to which two things are identical when they have all the same properties (in formal logic: $\forall F(Fx \leftrightarrow Fy) \rightarrow x = y$, where F is a property). To prove that mind and body are distinct substances, it is enough to point out at least one property they do not share. According to Descartes the great difference between the mind and the body is in divisibility. "[T]he body of its nature is endlessly divisible, but the mind completely indivisible." (Descartes, 2008, 61). He says that when he considers himself as a purely thinking thing (so probably without the modes of thinking that could be missing, like imagination or sensation), he cannot distinguish any part in himself and although the mind appears to be tightly connected to the body, removing some part of the body will remove nothing from the mind[7].

It is easy to give a sketch of counterarguments to both of the above arguments. In the first argument Descartes refers to the criterion of truth that he regularly uses in Meditations: clear and distinct perception of something. One way to diminish importance of this argument is therefore to say that "clear" and "distinct" in this criterion are metaphors, which beg for further explanation. The criterion is also entirely subjective, and a materialistic philosopher can always say that it only seems that mind can be conceived without a body. But the biggest problem of the argument in question is a non sequitur fallacy. Even if it is possible to conceive mind without body, it simply does not follow that they are distinct substances, even if for God it was possible to make them distinct substances. This is something that needs to be proved and Descartes does not do that. The second argument seems to fail in establishing Substance Dualism. It may succeed in showing that mind and body have different properties (or at least one different property), but it could still be the case that the properties of mind are properties of a composition of physical particles as a whole, not possessed by any of the particles by itself. So the argument is consistent with Property Dualism, but also other non-reductive stances in philosophy of mind, like Biological Naturalism (Searle, 2004). It is worth to mention what Rodriguez-Pereyra points out in his article Descartes's Substance Dualism and His Independence Conception of Substance (2008) – the fact that there are two different kinds of properties, mental and material, does not entail that there are two different substances. It could for example be the case that mental and material properties are distinct properties of

---

7 An interesting issue to consider would be implications of removing a part of the brain.

the same substance – this idea is called Property Dualism and it will be discussed in later course. Rodriguez-Pereyra also mentions real distinction between mind and body as a "weaker" possibility than Substance Dualism: the mind and the body could be numerically distinct substances which both have both mental and material properties. Substance dualism entails both Property Dualism and the real distinction between mind and body, but it is not entailed by any of these statements. Also Property Dualism and the real distinction between mind and body do not entail each other. We can imagine that mind and body are numerically distinct, but mental and material properties are the same and we can imagine that there is numerically one substance with different properties[8]. Another problem that is worth mentioning is how many numerically distinct individuals exemplify each substance – namely, are there many mental and material substances and they just share the same properties, or just one, etc.? In his article What Moves the Mind: an Excursion in Cartesian Dualism, Watson describes this problem as:

> [T]he theological demand that each human soul be an independent individual forces Descartes to posit an asymmetrical dualism, for while there need be only one material substance of which bodies are modifications, there must be a plurality of mental substances all of one kind. (1982, 73–74)

These arguments, counterarguments and issues, although important in specifying the nature of substance dualism, do not expose the most important problem associated with Cartesian dualism itself, at least the most important problem regarding the main topic of this thesis – action of an agent. The theoretical problem of whether there are two distinct substances, what their properties are, whether there are multiples of them or just one, etc., start to be of practical interest when consider the way they are related to each other. So far I have equated Cartesian dualism and substance dualism, but there is at least one important feature of Cartesian dualism that is not shared by every substance dualism flavour – namely, that the mind and the body interact with each other. In later course I will briefly discuss different dualist stances and it will be clear that the main

---

8   Or at least we can talk about these possibilities. Whether we can imagine them is debatable and could be itself an argument against Substance Dualism.

reason for their emergence is precisely the need to somehow answer the interaction problem that Cartesian dualism has. It is crucial for the main theme of this thesis, because it strongly influences what idea of agency can be defended. But before that, let us start with describing interaction of mind and body – the core idea of Cartesian dualism – and the reasons why it is so important – freedom.

Freedom of will

In the fourth mediation Descartes presents an interesting and influential view on freedom. According to him, human will is perfect (2008, 41). But by "perfect" he does not mean that it always makes the right choice. It is perfect, because its range is "unbounded by any limits" (2008, 41). For Descartes, the will could not be greater in any respect, and he claims that this could not be said by any other of human faculties, like remembering or imagining, that in God are "boundless." The interesting thing about the will is that Descartes claims it is incomparably greater in God, but yet it is perfect in a human being. The reason for that is that the will is greater in God in two respects: God's knowledge and power on one hand and the will's object on the other. We can think about it this way: perfect knowledge lets God always make the right choice and perfect power lets Him pursue the right thing under any circumstances. Also, the range of things that can be objects of God's will is greater than in case of human will (although it is probably caused by greater knowledge). Knowledge, power and range of things that can be an object of will are limited in a human being, but the will in itself is nevertheless of the same quality as in God. It is because will in itself is a faculty of having inclination or avoiding an object of will without being determined to have one of these attitudes by any external force.

The title of the fourth meditation is "Of Truth and Falsity," and it may be not immediately clear how we can draw a connection between the freedom of will and the truth and falsity but making sense of this connection will give us further insights into Cartesian conception of free will. Descartes recognizes in himself a faculty of judging and since it was given to him by God, he believes that if he uses it correctly, he will always get to correct conclusions. But if so, it would leave no room for doubt or error, and these are certain and common traits of the human process of gaining knowledge. What is even more challenging is that Descartes believes that it is impossible that God endow him with a faculty that is not perfect in its kind, which makes it clear to him that errors he occasionally makes cannot be caused by some imperfection in this faculty. The interesting explanation that Descartes gives for this problem is one involving another perfect faculty – will. Will's perfection makes it possible to want anything, even things that are irrational or contradictory to what faculty of judgement perceives as right. Freedom of will may be the reason why we sometimes make our decisions based on feelings instead of careful consideration of the things we want to decide about. This way, the faculty of judging is in us as perfect as possible and we still are capable of making errors (Descartes, 2008, 38–40).

Another important thing about the will itself, that is a result of above considerations, is that to make proper usage of our will and our faculty of judgement, we need to adjust our free will to our reliable judgement of the intellect, not the other way around, since we may want a much greater number of things than what we can understand (Descartes, 2008, 43). And the free choice consists of making a judgement and then conforming to that judgement with our will and act, even if the result of that judgement is not what we want. Descartes makes it clear, that the more we are compelled by our reasoning to one of the alternatives, the more freely we choose it. On the other hand, when we experience indifference, which is when we are not compelled by reason to any of the alternatives we choose from, we are still free, but it is what Descartes calls "the lowest degree of freedom," due to the fact that it is caused by the lack of knowledge. These remarks are of great importance, as it is easy to get into thinking of free will as a faculty that is completely undetermined and indifferent and to confuse being free with whatever one feels like doing.

Let us consider more closely the notion of free will endorsed by Descartes in context of the distinction of passions and volitions – the key distinction that is needed to understand agency in Cartesian dualism. It will also be extremely important in understanding why this notion of free will is rejected by materialists. The will that Descartes talks about comes in two general flavours. The first flavour is volition as a mental action and the second is the will's freedom as something distinct from it. As Jayasekera

(2016) argues, these two flavours are what underlies two reasons why Descartes assigns judgment to the will – judging requires something more than just perceptions and our ability to refrain from assenting to what we do not perceive clearly and distinctly. It is especially interesting that we can be in error even when we choose the true alternative. As Descartes says, if we do not perceive clearly and distinctly, then when we make a choice we are at fault no matter whether we choose the right thing, because we choose it purely by chance (Descartes, 2008, 43). One thing that binds together all these different aspects of will and is probably the most important to understand Descartes's view on it is the fact that Descartes believed that freedom of will is the only freedom we possess and only for our actions that stem from this kind of freedom we can be praised and blamed. It probably does not sound influential enough and therefore it needs some clarification. Actions here are mental and they involve what the soul can do without being affected by anything from the outside. And we can only be responsible for these actions, because everything else needs cooperation from the outside world. So the distinction here between our volitions and what follows from them is crucial. It is probably best expressed in his third maxim in A Discourse on the Method, where Descartes says that he should always try to master himself instead of the fortune (1912, 21–22), as we have no control over what happens after we make a particular decision.

We can already see here a crucial difference between Descartes's account and for example the kind of control compatibilists talk about. It is hard to imagine that a compatibilist would say that a person has free will when they can decide to do something and then not be able to execute it. In the core of compatibilism is the idea that a person is free when they can do X if they decide to do X. When Descartes says about being free to choose X, he rather means controlling our inclinations towards X. The consequences of this are tremendous, although probably not visible at first glance. In Dennett's compatibilism people are responsible for their thoughts only when the consequences of their deeds are disadvantageous. For example, if we have two people with exactly the same brains who make the same decision in a similar situation, only one whose actions result in outcomes that are socially unacceptable will be punished with an aim of changing their behaviour in the future. In Descartes's view, if there are two people that make the same decision having the same reasons to make it, i.e. when they assent to what they clearly and distinctly perceive or when they withdraw their assent to what they do not perceive this way, it does not matter whether the consequences of their deeds are different – they are equally responsible, as we cannot be held responsible for what we cannot control. Descartes's view on free will also changes how we can attribute praise and blame. Since people cannot control what they accomplish, they should not be held responsible for it. Things we can be held responsible for are always relative to the situation we find ourselves in. The peculiar consequence of this view is that a person who did not accomplish anything, but always made his/her judgment compliant to what they clearly and distinctly perceive is more praiseworthy than a person that took a risk that eventually let them accomplish something that would be otherwise unachievable.

All these remarks are important to get to the conclusion why Descartes needs an immaterial soul to establish his version of freedom of will. It is because only this way the will can be free from any external influences. Emotions, for example, are also a part of us, but they are tightly connected to our physiology. Controlling our emotions means something completely different than controlling what we assent to, as due to connection to our physiology emotions are something not fully dependent on us. It is rather that even though we have emotional attitudes that incline us to something, we have to control what we assent to – if we did not have these attitudes, there would be at least one fewer concern for us. It is hard to imagine how Descartes's conception of responsibility could be saved if there were no immaterial soul assumed. To make it more concrete, I will draw a connection to one of more contemporary accounts of free will. Eccles tried to argue that denial of free will is neither rational, nor logical. His argument reads:

> This denial either presupposes free will for the deliberately chosen response in making that denial, which is a contradiction, or else it is merely the automatic response of a nervous system built by genetic coding and molded by conditioning. (1976, 101)

There is one problem with this response – it seems to assume that free will is just deliberately choosing something and that it is not possible for an automatic response to be deliberate, or it just needs a more extensive justification of free will. But at any rate, we can try to get a gist of it. There is no reason to call anything free will if the only response that our bodies can make is based on built in or conditioned reactions of our nervous system to stimuli. For if it was so, to get back to Descartes's terminology, our assent to what we clearly and distinctly perceive would also be determined by build in or conditioned reactions of our nervous system to stimuli. Moreover, even the idea of perceiving something clearly and distinctly could be questioned as determined by our nervous system, i.e. I could never know whether I perceive something clearly and distinctly or it is just my nervous system that makes me think so – my nervous system would be Descartes's Evil Demon. And Descartes does not seem to have any solution that would not preclude existence of free will if the will were at least partially determined by the matter, let alone if it were a function of material causes.

As we could see, it is human responsibility that is one of the main reasons why Descartes needs two substances. Since material substance is separate from the matter, we can have control of what we think, but one piece is needed to establish at least some control over what we do – how mind and body coexist with each other and what is the way they influence each other. Substance dualism endorsed by Descartes solves this issue by introducing interaction between mind and body.

What would happen to responsibility if there were no interaction between mind and body in Cartesian dualism? From what we have seen in the previous subchapter, one possible answer is "nothing," as we are responsible only for our thoughts. But it is hard to imagine that Descartes would agree for that. We are not responsible for what we accomplish because we do not have control over world's cooperation, but it does not mean that if our thoughts were totally disconnected from matter we would still be able to exercise the same kind of freedom. For example, we can think of a paralyzed person and their freedom. From Descartes's account we can derive that this person would have still the same freedom of will, but it is rational to expect that the fact that they know they cannot exercise their freedom in the physical world would influence the way they exercise their freedom at all. After all, it is always easier to conform to the rules one has to conform to only theoretically – it is completely different when one knows their decision may have impact in the physical world, even if it is in principle impossible to control it.

Interaction between mind and body is also needed to account for a human being as a composite individual – composed of mental and material substances at the same time. It is probably prevalent to interpret Cartesian dualism as a theory in which mind and body are only remotely related and underestimating the role of the body in composing a person. But Descartes makes it clear that he considers himself a being composed of a mind and a body, rather than a mind to which body is just somehow attached. And without interaction between mind and body even such a simple idea as behaviour caused (at least partially) by emotions could not be introduced. At any rate, Descartes needs interaction between mind and body in his solution to the psychophysical problem and by considering them as thoroughly distinct, he makes it easy to make numerous objections to this idea. Let us now get through them briefly and then discuss them in greater detail.

Problems of interactionism in Cartesian dualism

There are three main issues that Cartesian Dualism has to face with respect to the interaction between mind and body I would like to discuss. First of them will be the place where this interaction takes place, second the problem of causal closure principle and third the problem of conservation of energy.

The problem of the connection between mind and body is an issue widely discussed concerning Cartesian Dualism. According to Descartes the mind interacts with the body through the pineal gland and of course it is a very problematic statement. But first of all it is advisable to think about the reason why Descartes needed to find one place in the brain (or the whole body) where the interaction between mind and body takes place. Before he talked about the interaction in the pineal gland in The Passions of the Soul, he embraced the view held before him by St Augustine or St Thomas Aquinas, that soul is attached to the whole body, not one particular part. But, as Lokhorst (2013) rightly notices, he rejected an important idea that was associated with this view and accepted by Augustine and Aquinas, namely, that the soul is the principle of life and what happens in the body is explained not only using materialistic concepts, but also referring to the soul – for example beating of the heart or the reception by the external sense organs of light were explained by concepts of vegetative and sensitive soul. Descartes seemed even to be proud to say that we can get rid of vegetative or sensitive soul while considering processes that take place in the body. But with introduction of two interacting substances Descartes introduced also new difficult problem to resolve. It is not easy to find a solution to the question "how a soul can be at the same time attached to the whole body, but interact with it through the pineal gland?" In order to do that, Descartes makes a comparison between the soul and gravity. When a body hangs from a rope attached to any part of it, it pulls the rope with all its force, as if all the gravity existed in the part that touches the rope, while it is actually spread through the whole body.

Even if the comparison between gravity and soul is sensible, there is still a problem of how an immaterial soul can interact with material pineal gland. According to Lokhorst (2013), there are a few reasons why Descartes considered the pineal gland as a place of the interaction between mind and body. Pineal gland is the only single organ in the brain, it is also small, light and easily movable. First of all, none of these reasons seems convincing – for example it is imaginable that there are two parts of the brain to which one soul is connected. Secondly, and more importantly, these reasons do not give any answer to the question how the interaction occurs. It is hard to believe that any conclusive answer can be given to this problem. If mind and body are radically different, we can probably only talk about correlations between states of body and states of mind.

I believe this is actually the least important problem of Cartesian Dualism. If it is not possible to establish existence of mind based on existence of body or body based on existence of mind, then at least the distinction between mind and body has to be acknowledged as real and, as I have already stated, we can only investigate correlations between mental and physical events. The fact that there is a problem to show where and how the mind is connected to the body is not something that can disprove Cartesian Dualism in general. Of course, Cartesian solution could be viewed as an ad hoc manoeuvre to cover some difficulties of Descartes's theory that are not possible to overcome but saying that pineal gland has other responsibilities in human body than accounting for the connection between mind and brain is not a good argument against Cartesian dualism itself. In Meditations Descartes considered various problems concerning philosophy of mind and some may require dualistic solutions.

Another problem of Cartesian Dualism and especially Interactionism is a Causal Closure principle. Kim (1998, 40) defines Causal Closure in this way: "If you pick any physical event and trace out its causal ancestry or posteriority, that will never take you outside the physical domain. That is, no causal chain will ever cross the boundary between the physical and the nonphysical." To be more specific and to apply this principle to our area of interest, no brain event has any immaterial cause. As Wachter (2006) notices, it is possible to restate this statement in two versions: weaker and modal. A weaker version of the Causal Closure principle would be to say that if a material event has a cause, then that cause is wholly material. Wachter states modal version as "it is impossible that a material event has an immaterial cause." So, in other words, in every possible world, every material event has a material cause. It is important to notice that distinguishing the weaker and the modal version of Causal Closure principle means that the weaker version must refer only to material entities in the actual world – otherwise stating the modal version would be redundant and the weaker version would not be weaker than what Kim stated, after all. This is important, because it may already show one of the problems with the Causal Closure principle: what makes our world "special" with respect to other possible worlds that we can exclude a possibility for any event to have an immaterial cause?

If Descartes is right and there is interaction between mind and body, then there must exist some physical events that have immaterial causes – even if we take it for granted that interaction between mind and body happens in the pineal gland, it is still the case that mind, an immaterial substance, needs to be a cause of some changes in the pineal gland. But the Causal Closure principle is more general and refers not only to the mind-body problem. It makes any kind of immaterial object causally ineffective. And it is not hard to find an example of an entity that is not a human mind, to which at least some of us ascribe efficacy: God's causal influence on material world must be recognized as impossible if someone assumes that the Causal Closure principle is valid. In other words, the material world is a world without miracles, since any kind of divine intervention must be ruled out.

Is Causal Closure principle valid and what are the reasons to believe so? Wachter (2006) in his paper *Why the Arguments from Causal Closure*

Against Existence of Immaterial Things is Bad analyses at least some possible answers to the second part of this question and it is advisable to mention here his results. According to Wachter, there are two general arguments for the Causal Closure principle: (1) it is presupposed by science or rationality, (2) it is supported by the success of science. He claims that none of these statements is true and here I will consider why.

If Causal Closure principle is assumed by science or rationality, there should be some good reasons to do that. Wachter mentions an utterance formulated by Kim to support this assumption. According to Kim, rejecting Causal Closure principle is the same as, or at least entails, rejecting completability of physics. It is quite obvious and can be treated as analytic *a priori* judgement based on what Causal Closure principle is and what it means to reject it. Unless we introduce other, more "basic" reality than the physical one, that could interact with physics and yet be tractable by science, rejecting Causal Closure principle means that there are some events that cannot be explained using scientific methods. But Kim also adds that no serious physicalist could accept this result. It is probably quite arbitrary whether somebody wants to be a serious physicalist or not, but this idea gives some clue what the issue with rejecting the Causal Closure principle is. What would be lost if we rejected this principle?

As Wachter seems to answer, "serious physicalists" may be worried about validity of the laws of nature that we would discover in a world in which some physical events could have some immaterial causes. In other words, the question is whether what we call "laws of nature" are really laws of nature if we reject the Causal Closure principle. According to Wachter, there are three ways in which laws of nature can be related to the principle in question. Counterinstances to the Causal Closure principle can violate the laws and therefore falsify them, violate the laws without falsifying them or not violate the laws and of course not falsify them at the same time. Which result can be accepted depends on what notion of laws of nature we accept.

In "Humean" vision of the laws of nature, as they are presented in *An Enquiry Concerning Human Understanding* (Hume, 2007), any violation of a presumed law of nature falsifies it. According to Hume, since "firm and unalterable experience" has established laws of nature, we have as strong evidence against miracles as we can get from experience. In other words, an event is treated or thought of as a miracle when this kind of event has never occurred before. Of course one might ask a question whether any case when something immaterial moves something material is a miracle, but since immaterial causes violate completeness of physics, they can probably be treated as miracles that happen regularly due to how human beings are formed (if Substance Dualism or actually any kind of Dualism is true).

But there are also other views of laws of nature that let us make sense of how miracles can happen without falsifying the laws of nature. One of them is the way Richard Swinburne treats miracles as violations of laws of nature, which is presented by Wachter as an example for how miracles can be counterinstances to the Causal Closure principle and therefore violate the laws, but yet not falsify them. According to Swinburne, if we formulate a law of nature L that we have a lot of evidence for and that explains many experimental data, is simple and fits with the other laws, we may have good reasons to keep it as a law even if a counterinstance to L occurred. The reason for that may be the fact that adapting L to take account of the counterinstance may lead to a new law that would be more complicated or yield many wrong predictions, probably on the data points for which L worked fine. In such case the counterinstance should be treated as a non-repeatable counterinstance to L.

The third option given by Wachter is to oppose Humean laws of nature to the way Mill talked about tendencies in nature. Humean laws entail regularities: some events are always followed by other events. According to Mill, laws of causation should be stated in words affirmative of tendencies, not of actual results. Wachter explains what it means using as an example the law of gravity. As he claims, the law states that in a situation of a certain kind there is a force of a certain kind, as opposed to a law that would state that bodies situated in a certain way will move in a certain way. This is an important distinction if we want to account for forces that may act on the bodies, which is actually explicitly accounted for in the definition of the law of gravity, which states that bodies will accelerate in a certain way if nothing else is acting on them. As Wachter notices, in many cases in which the law applies, there is no claim about how the body will actually move. This entails that laws do not entail regularities as they were understood by Hume. An objection to this, which Wachter brings himself, is that although a single law does not entail regularities, all laws together do. In other words, the phrase "if nothing else is acting on them" refers to other physical entities whose influence on the moving entities can be spotted by science. But it is not so simple, since to every prediction based on the laws of nature we have to add the clause "if nothing else is acting on the things in question" or similar and it does not state what are the things that may act. A physicalist will of course exclude God or other immaterial entities that could act on the "things in question," but this exclusion bears a question. Laws of nature simply do not state which things exist and which do not. In this view on the laws of nature if an immaterial entity like God acts on a material entity whose movement cannot be explained using laws of nature, it is not any violation of these laws. As Wachter says, if God moved the stone from Jesus' grave away, He did not have to abolish any forces, he "overrode" them[9].

According to Wachter, the last view of the laws of nature is the correct one and he tries to answer the question whether it is presumed by the science that the material world is a causally closed system assuming that the notion of the laws of nature in question is correct. As he rightly notices, science is usually not concerned with scientific explanations of single events, rather with general questions. Science is interested in regularities and therefore it is not important whether sometimes material objects cause material events. It is important only to ensure, that this is not the case during experiments that try to establish physical relations between material things. In cases when science however is interested in looking for law-based explanations of particular events, it is also not a problem if some immaterial objects cause material events, because science may look for law-based explanations, it will just not find them, because they do not exist. If science could not find explanations of some material events, like brain activations, it would actually be evidence for existence of some immaterial objects.

---

9   Of course one may ask whether there is any substantial difference between abolishing laws and overriding them. Wachter does not discuss this issue in his article. Intuitively when something is overridden, then it ceases to exist, and a new thing takes its place. However, I believe that this remark does not influence the main point of Wachter's argument – that laws of nature are not violated by immaterial entities acting on material ones.

There is something important here to notice. Wachter claims that due to the fact that science is usually concerned with general questions, the events of God moving a stone or souls causing material events in human brains do not impede discovering laws governing stones and elementary particles. He is right about that, but he misses an important point. If there are souls that interact with human brains and cause events in them, then it is crucial how this interaction happens. God does not move stones every day, but if souls exist, they are probably constantly attached to bodies, or brains, or whatever they are attached to. It is hard to imagine how this may not cause any troubles in formulating laws governing how brain works. Of course it is not a problem, at least in principle, to establish activation of various parts of brain given that other parts of brain are activated. But if souls constantly interact with brains, there is some randomness constantly involved. There is no starting point in which a soul interacts with a brain and causes a chain of brain activations to happen. Even if the place where soul interacts with a brain is isolated (in other words not the whole brain is connected to a soul), there is still some indirect influence of soul on other parts of the brain. Therefore the validity of the idea of dualism is based on the fact that we cannot predict accurately activations of some parts of the brain given activations of other parts of the brain. If in the future we will be able to do this accurately for any part of the brain, then the whole concept of the soul interacting with the brain will be impossible. But it is a big problem for Dualism even now, because it is based on present lack of knowledge, which not only may be temporary, but also should not be a basis for a philosophical stance. It resembles a situation when a theist tries to support his/her belief in the existence of God by finding gaps which cannot be explained by science (at least for now) and claiming that they can be explained by the existence of God. It has also practical consequences. A belief in the existence of the soul that interacts with the brain is an important part of the worldview of many people and a brain scientist who believes in dualism would therefore have hard time trying to discover relations between parts of the brain that could undermine his own worldview.

But let us get back to the issue of the Causal Closure principle. According to Wachter, the principle in question is not presumed by science as it was presented above. The second question is whether the success of science supports the Causal Closure principle. Here Wachter also gives a negative answer. As he says, "science has made many successful predictions and is likely to do so in the future" and "science has discovered laws of nature and is likely to do so in the future" (Wachter, 2006, 10). But these two statements do not support the Causal Closure principle, since they only show that in many causal processes there were no immaterial things that intervened. There are events for which we cannot make clear predictions based on laws of physics and it supports an idea, that there at least may be some immaterial entities at play. There were also individual events in the past whose causes were not investigated by science and science will never investigate them and some of these events could also be caused by immaterial entities. As Wachter rightly points out, this is hardly evidence for the existence of immaterial entities – we should believe in them if there is evidence for them – but it shows that it is possible for immaterial entities to coexist with the success of science.

Once again the last argument is an argument based mainly on the gaps in science which could be filled with a hope that there are some immaterial entities at play. Wachter is right about the point he is trying to make, but he does not make an argument for falsity of the Causal Closure principle. However, he gives an argument against using induction in proving the principle, which is worth mentioning here, because it shows flaws in the reasoning that are common on the physicalists' side. He mentions Melnyk, who argues for materialism using the Causal Closure principle. According to Melnyk, since science has found sufficient physical causes for physical effects of many kinds, we should conclude by induction that the Causal Closure principle is true. But by induction we can only infer that when we discover forces existing in a particular case, there are also such forces in all situations of the same kind. But we cannot argue by induction that the Causal Closure principle applies in all kinds of situations. To argue for this by induction would mean to take all kinds of situations in which the Causal Closure principle holds and project it to all other kinds of situations, but it is not a proper usage of induction. Based on the fact that often two things accelerate towards each other we cannot argue that two things always accelerate towards each other and based on the fact that often material events have a full material cause we cannot argue that it is always the case. There is no way to argue for materialism or naturalism from the fact that assumptions of materialism or naturalism work well in certain situations. This is actually important in the context of what Wachter tries to establish. He does not try to convince anybody that there are certain immaterial beings at play in different situations, but that we cannot use the Causal Closure principle to prove that there are no such entities. But, once again, it does not help Dualism too much. Dualism will always be a weak stance if it is based on the contemporary stage of scientific knowledge. And the arguments against the Causal Closure principle discussed above seem to be of this kind. Since they are based on the contemporary stage of scientific knowledge, they may be only temporarily interesting.

A different argument is given by Swinburne, one of the most prominent contemporary defenders of Dualism. He claims that no experimental result of science can justify the Causal Closure principle (Swinburne, 2013, 117). The problem with such a justification is that any experiment that involves showing that conscious events do not have any causal influence on the physical world must use testimony of agents that experienced these conscious events. For example, when a scientist wants to show that an intention did not have any influence on whether a subject of an experiment raised their hand, they need to know that this intention occurred and the only way to know about it is to ask the subject about it and assume that what the subject said was caused by the experience of an intention. And this already violates the Causal Closure principle (Swinburne, 2013, 119). Another option that Swinburne takes into consideration is to approach the problem from the other side – instead of showing that conscious events do not have any influence on the physical world by picking a conscious event and showing that it does not have any influence on the physical world, it may be more tractable to show that everything that happens in the brain has other brain event as its immediate necessary and sufficient condition. As Swinburne argues, someone could justifiably believe that some brain events are caused solely by other brain events only in case of their present observation. It is not possible in case of past experiences of causal closure of the brain, because if the brain is causally closed, then a memory of a past experience could not have been caused by that experience and therefore loses its credibility (Swinburne, 2013, 121). In general, no matter what we try to do to justify the Causal Closure principle, we always fail because the truth of the Causal Closure principle would mean that our conscious events had no part in forming the justification.

To be able to get to this conclusion Swinburne of course must have a reason why to think that conscious events never have any part in forming a justification if the physical universe is causally closed. And Swinburne has that reason, and it is grounded in the ontology he defends – that there are pure mental properties whose instantiation does not depend on instantiation of any physical property (Swinburne, 2013, 68). Mental and physical properties are distinguished using a criterion of whether they can be publicly accessed (physical) or the person that has them, has privileged access to them (mental). Privileged access to a property for a person means that whatever ways others have to access that property, there is always another way in which that person can access it and this way is not accessible to others. In this way the stance Swinburne defends becomes similar to old good Cartesianism and unfortunately inherits its problems that I will not mention here again. But at any rate, Swinburne shows that arguing for the Causal Closure principle is not an obvious and straightforward thing as more physicalistically oriented philosophers can think.

Very similar to the problem of compatibility with the Causal Closure principle is the problem of compatibility between Cartesian dualism and conservation laws. The main idea is that interaction between mental and material substances (or properties) is impossible, because the amount of energy in the universe is constant and causal interaction of mind and matter would have to either result in reducing the amount of the energy in the physical world (when matter acts on the mind) or in increasing the amount of the energy in the physical world (when mind acts on the matter). There are philosophers that believe this is a decisive argument against Cartesian dualism. For example Dennett claims that the principle of the conservation of energy is fundamental and standard for physics (1991, 35). I will not discuss this issue here in details as I did with the causal closure principle. There are philosophers that argue the problems with the principle of the conservation of the energy stem from misunderstanding – the physical view on the conservation of energy is much more complex than that it just works in every case for every physical entity. For example, one detail that is often omitted when somebody wants to use the energy conservation principle against dualism is that the principle does not hold for the universe as a whole, but for each place in the universe separately. Pitts in his dense article *Conservation Laws and the Philosophy of Mind: Opening the Black Box, Finding a Mirror* (2019) discusses this and many other responses to the problem. One of the most important conclusions, as in case of the causal closure principle, is that if somebody believes in this kind of physical arguments against dualism, they already presuppose materialism.

Problems of Cartesian dualism with respect to free will problem

Let us now think how these issues and Cartesian dualism itself can be related to the problem of autonomy discussed in this thesis. Here, we can look at the problem from two perspectives. First, the perspective of counterarguments to Cartesian dualism and whether by undermining Cartesian dualism they also undermine free will (how it is defined by Descartes) and second, the perspective of the notion of free will within Cartesian dualism itself – whether we can derive a coherent notion of autonomy from it or not.

Let us start with the discussed problems. How severely do they impact the notion of free will as Descartes understands it? I would say that they are rather inherent to Cartesian dualism as a stance that is supposed to solve the psychophysical problem and they affect the Cartesian idea of free will indirectly, by affecting Cartesian dualism. It means that of course they may severely affect the reasons to believe in Cartesian dualism, but the consistency of the idea of free will is not affected. The specific way how mind and body are related to each other makes no difference to the idea of acting freely. As I have already mentioned, it is disputable whether any kind of connection or interaction can influence free will as it is endorsed by Descartes – if we are responsible only for our decisions, not what we accomplish and this is because we cannot control world's cooperation in our accomplishments, then even a total lack of control over our bodies would not diminish our freedom. It may sound counterintuitive, but this seem to be a result of Descartes's concept of free will. If we leave aside this extreme case and assume that an interaction between mind and body is needed for freedom, the fact that it is not clear how this interaction might happen may make it harder to believe this kind of freedom is possible. Probably this is the main problem with Cartesian dualism in relation to free will, it is not about the coherence of the whole stance, but rather about reasons to believe it is true. Similarly, the arguments against Cartesian dualism derived from physics cannot affect the Descartes's idea of free will, but they are supposed to make the belief in Cartesian dualism less rational or even irrational. But as we could see, these arguments from physics seem to stem from a bias of philosophers towards wrongly understood physics and they are far from being conclusive.

The problems of the account of free will endorsed by Descartes are rather similar to those of incompatibilism. If someone believes that the Rollback Argument is a strong argument against compatibility of free will and indeterminism, the idea of making decisions while being isolated from the influence of nature may also make impression that the resulting decisions are random. On the other hand, the type of autonomy that Descartes gives to the soul makes it quite easy to employ the notion of agent causation in explanation of where the source of the decision is. Actually, this kind of agent causation makes it easy to also explain why decisions are free even if they are determined by reasoning (or the more they are determined by reasoning, the more free they are). By now it should already be visible, that a stance like Descartes's cannot be easily categorized as incompatibilism or compatibilism. On one hand it is clear that soul's decisions are not a subject of determination from nature, otherwise they would not be free. On the other they are a subject of another kind of determination – reasoning.

It is of course a serious problem to explain how a human being that is composed of mind and matter can choose between these two kinds of determination – after all, if it is determined that he/she always chooses reasoning over feelings or emotions, then we might have a problem in explaining why we call an action free. Descartes seems to make some room for indeterminism in human choices, since he says it is not necessary that he can always choose both alternatives to be free. But there are solutions like Leibniz's, where he makes it clear that all human actions are determined and free at the same time and to explain how it is possible he makes an extensive use of agent causation. It may be tempting to call Leibniz's stance a compatibilism[10], but we have to remember that in his determinism a conscious substance determines itself in every respect – there is no causally efficacious influence from nature. It is quite different than compatibilism where it has to be explained how decisions can be free when they are results of natural processes that are outside the person that makes a choice.

Another problem with the notion of free will in Cartesian dualism is that the notion of control is not entirely clear. It is also quite different than what compatibilists or incompatibilist have in mind when they talk about control. It is quite clear that for example for Van Inwagen to have control over something means to be able to make it happen in the physical world. Both compatibilists and incompatibilists would say that to have control over something an agent must be able to do x when he/she wants to do x. In case of Descartes the control happens on, we can say, metalevel, since the control we have is a control over our decisions, not about their outcomes. But, as I have already mentioned, it is hard to believe that a person that cannot influence the physical world would treat his/her control in the same way as a person that knows that their decisions can be effective.

Here I will stop my initial remarks about dualism in general and Cartesian dualism in particular, as it is enough to get an idea of its role in the free will debate – we will also learn more about the dualist view on free will when we compare it with materialist stances. Let us now briefly talk about other dualist stances, focussing on the differences that are supposed to help in better understanding of agency.

Other dualist solutions

A natural next choice after discussing substance dualism is property dualism, whose aim is to make it easier to explain the relation between the material and immaterial substances. One of the most famous flavours of property dualism is expressed in the Chalmers's (1997) book *The Conscious Mind*. Already in the preface it is clear that Chalmers's approach is quite different than Descartes's. He speaks about consciousness rather

---

10 This is how for example Jorati (2017) describes Leibniz's stance.

than about a soul (quite ironically, the word "soul" appears in the book twice, only in titles of other books, one of them being the title of a book of an eliminative materialist, Paul Churchland – *The Engine of Reason, the Seat of the Soul* – whose view on consciousness I will discuss in the next chapter) and his thinking about the consciousness already fits into the paradigm where the consciousness is supposed to be explained by science. He believes there are good reasons to believe that physical systems like brains are what causes existence of consciousness, yet it is not clear how it happens that a brain is not only a physical object, but can also experience things (Chalmers, 1997, ix). According to property dualism there is only one kind of substance in the universe – physical substance. But there are two kinds of properties: physical and mental. What does it mean in practice? It means that there are events in the universe that cannot be explained using only physical properties of physical entities. Following Robinson (2016), we can take a hurricane as an example of an entity whose existence can be explained using only physical properties of the matter it consists of. Namely, hurricane is identical to the atoms it is composed of, but we cannot give one physical description that would apply to any hurricane we encounter in the world. This is usually referred to as token identity as opposed to type identity. According to property dualism there is no token identity between atoms a person consists of and his or her mental experiences. The important difference between property dualism and substance dualism is in the existence of the carrier of mental properties. In case of substance dualism there is an external carrier, namely a separate substance, while in case of property dualism the individual that we ascribe mental properties to is the body of a person.

The most important from the perspective of the main topic of this thesis is what the new possibilities are when we assume property dualism instead of Cartesian dualism. As Lycan tried to prove (2013), property dualism is not much better off (if at all) than Cartesian (or any kind of substance) dualism. Instead of having difficulty with understanding how mind and body can interact with each other, we have a similar difficulty with understanding how a material brain can bring into being immaterial properties. If we accept the view that immaterial properties can interact with material properties, the problem of consistency with physics (if we agree there is such a problem) also retains.

Churchland (2013, 33) claims that property dualism makes it easier to account for degenerated activity of mind when brain is damaged, because it reckons physical brain to be the seat of all mental activity. It is however not clear why a perceived activity of an immaterial substance could not be in a similar way dependent on the physical body to which it is connected. We have to remember that there may be a difference between what we perceive as a result of thinking in the physical world and the thinking itself. After all, from Cartesian perspective body is just a biological machine and it is not hard to give examples of machines that malfunction so badly that even the best operator cannot do anything about it. It may be considered an *ad hoc* manoeuvrer (probably by a materialist), but I see no reason why a Cartesian dualist could not choose this path. Also, according to Lycan (2013), it is possible to a substance dualist to accommodate detailed dependence of mind on neural activity. As he reminds, Descartes makes an exception for free will and this way distinguishes humans from machines[11].

What are the advantages of property dualism when we tackle the free will problem? Let as answer this question using divide and conquer strategy. First we will split property dualism into two theories based on the fact whether they assume that conscious, immaterial properties are causally relevant in the physical world. If we assume they do not, we arrived at epiphenomenalism, one of the flavours of property dualism. It is a stance that although mental properties do exist, they are causally impotent – their existence cannot change anything in the external world. It is important to notice, what is correctly identified by Lycan (2009, 557), that it is not a stance that helps to completely bypass the mind-body problem interaction at all – interaction still happens, just in one direction, from body to mind. What it can help with is to avoid problems with the violation of the causal closure or energy conservation principles while retaining some kind of reality of phenomenal experience.

---

11 Lycan also says a dualist does not have to accept free will to accept substance dualism, but it is not a very interesting move from the perspective of the main topic of this thesis.

It seems to be clear that epiphenomenalism does not let us say anything sensible about free will except for that it does not exist, but some philosophers try to make an argument for the opposite conclusion. For example, according to Nadine Elzein, it is not clear why causal irrelevance of mental phenomena must lead to any conclusion about free will. In her article *Free Will & Empirical Arguments for Epiphenomenalism* (2020) she makes her point in the context of contemporary research in neurobiology and additionally makes a claim that only adopting highly non-naturalistic assumptions about the mind like Cartesian dualism may be threatened by neurobiological research, which makes a perfect case for us to investigate.

According to Elzein philosophers are surprisingly not interested in pursuing the problem of relation between free will and supposed impotence of mental states (Elzein, 2020, 3). I would argue that it is in general surprising that contemporary philosophers pay not enough attention to the relation between the free will and psychophysical problems in general, not only in this case, and therefore we have so many general, but not very useful, arguments for different cases as ones presented in the first chapter of this thesis. Elzein proceeds with a claim that on the other hand neuroscientists' research (like for example Libet's experiments (1979)) is interesting primarily because of the case that it is devastating for free will. She forms three possible explanations of this disparity. First, that neuroscientists are right and their case for epiphenomenalism has more serious implications than philosophical arguments. Second, that philosophers are just oblivious to the threat that epiphenomenalism poses to free will. And third, that the empirical researchers are mistaken, there is no problem of compatibility of epiphenomenalism and free will (so there is nothing that philosophers should care about). Elzein of course supports the third option.

Elzein first splits conditions of freedom into two: naturalistic and non-naturalistic. Let us start with naturalistic conditions for freedom. She mentions four of them: acting on the basis of choices, responsiveness to reasons, harmony with deeper values and alternative possibilities. All of them have a similar response, so I will not discuss all of them in detail. Of course, Elzein claims that epiphenomenalism poses no threat to naturalistic conditions of free will. Why? Because none of them makes it explicit how relation between consciousness and the fulfilment of the condition should look. For example, one may argue that if epiphenomenalism is true, then our reasoning may have no impact on what we do – after all, what our body does is determined by physics. But here, of course, definitions come to the rescue. Elzein points out that our choices do not have to be initiated consciously or that they have to be efficacious in virtue of conscious features (2020, 11). She says that it is not crucial that our awareness of reasons must be causally efficacious, it is enough that there is correspondence between the processes we are conscious about and rational consistency. She makes it explicit that even if these processes were initiated by unconscious events, they could still be governed by these norms, and this is what is the most important. The general strategy seems to follow a pattern "y is a condition for freedom and maybe consciousness has no causal influence on y, but y can still be met due to x, so there is no threat to free will."

As the first non-naturalistic condition for freedom Elzein mentions conscious origination. She claims if we assume causal influence of consciousness is needed for a free action, then probably neuroscience gives us good reasons to believe freedom is not possible. But conscious influence is a way to preserve agency and it still may be preserved if we on the other hand assume that unconscious processes in the brain are part of the agent. The second non-naturalistic condition she talks about is immunity from prior influence. Elzein makes a point that here a correlation between agent's choices and neural events that precede these choices can be exploited. It is possible that a pattern of neural activity is associated with positively assessing some course of action – but it does not determine it. The positive assessment of a course of action is of course correlated with having a reason to choose that course of action and this way neural event is correlated with having reasons to choose some course of action (Elzein, 2020, 14). Once again, if we insist that rational explanation must have its origin solely in the soul of an agent to count, then there is a reason to doubt in free will.

There are two things to say about these deliberations. One of the mistakes that Elzein does is that she mistakenly associates Cartesian dualism with some vulnerability to arguments from neuroscience, while other stances are allegedly immune to these arguments. It is not that only Cartesian dualism is threatened by advancements in neurobiology (if it is at all), it is rather that Cartesian dualism enables some idea of free will that is impossible to make sense of on the ground of some other stances, like epiphenomenalism and materialism. It is absurd to say that free will is threatened only if we adopt non-naturalistic assumptions about the mind, like Cartesianism, whereas other stances are immune to development of neuroscience and free will is completely fine and well on the ground of these stances. It is absurd to say that, because words "free will" refer to a completely different concept when we use them on the ground of Cartesian dualism and when we use them on the ground of epiphenomenalism (if somebody insists on calling this suspicious phenomenon free will, but as I said at the end of the previous chapter, I am not going to argue about definition choices). It is enough to say that the free will that Elzein defends would also be completely defensible on the ground of Cartesian dualism. But if Descartes wanted to defend this notion of free will, he would not need his dualism, at least in this respect. The arguments that Elzein makes have a form "x is defended by materialists and, it is also possible to defend it on the ground of epiphenomenalism" without any critique of whether x is a sensible in the first place. It is important to notice, because it means that compatibilism may become a go-to stance whenever there are problems with free will on the ground of a philosophical theory.

Another problem is that if we assume materialism gives a sensible foundation for free will (more on that in the next chapter) it is not clear that epiphenomenalism can just transfer materialistic arguments to its ground. We have to remember that on the ground of materialism also consciousness is explained in physical terms and therefore saying that, for example, something was up to somebody even though it was determined by the laws of physics may at least sound more plausible – the conscious processes are supposed to be material, so it is not a problem for them to be efficacious. But here we have to drop the idea of causal efficacy of consciousness, but still pretend there is some sensible notion of self that can help us explain why it was up to us that we did something. It is especially visible when we consider how Elzein explains acting with reasons.

I believe the problems presented here are typical of epiphenomenalism and even a short assessment of epiphenomenalism is a good prelude to the next chapter, where we will see how philosophers change the notion of free will to make it compatible with materialism. As I have already indicated, there are some issues of epiphenomenalism that materialism can at least mitigate, but as we will see, the conditions of free will are for materialists similar to what was mentioned by Elzein.

In the end, if epiphenomenalism is true, it is hard to make sense of any notion of free will that would be attractive in any way. On the ground of epiphenomenalism on one hand we lose strong notion of mind proposed by Descartes, on the other we lose any causal influence of consciousness that can be retained on the ground of materialism. Now, is any other idea of property dualism in better position to explain free will?

It is possible to claim that mental states produced by brain states are causally efficacious thanks to the brain properties they are embedded in or on its own. Let us now investigate these two options. First, what advantages would it have to assume there are mental properties, but no substance, which have causal powers to influence the matter? I believe the main reason (or at least one of them) why Chalmers wanted to introduce property dualism was to be closer to naturalistic science and according to him property dualism is a naturalistic dualism that is supposed not to violate principles of physics like the causal closure principle (Chalmers, 1997, 11). I have already referred to the alleged incompatibilities of dualism and physics and in this respect Chalmers seems to be one of the philosophers that are guilty of trying to make his ideas compatible with science even though there is no incompatibility found. But if we take this goal seriously, is property dualism in "better" position than Cartesian dualism? It may be claimed that it is easier to reconcile it with the theory of evolution, but on the other hand if immaterial properties were brought to existence by some kind of physical processes (which must be the case in Chalmers's perspective) and we have no idea how it exactly happens, we can as well say that immaterial substances emerge from matter in course of evolution or that God at some point attaches immaterial substance to some material substance in course of evolution[12]. And except for the fact that for some people it may be an advantage of property dualism that it does not assume existence of immaterial self drastically separate from material substance it is hard to see any advantage of property dualism over Cartesian dualism. And it is impossible to retain the notion of morality used by Descartes and

---

[12] An idea that Cartesian selves can show up in the process of evolution is mentioned for example by Lycan (2009, 10–11).

associated with his idea of free will if we reduce the self to mental properties emerging from physical substance.

As I said, it is also possible to defend causal efficacy of mental states due to physical states they are embedded in. It is however hard to say whether such a view would be a property dualism anymore – it is rather a form of non-reductive physicalism. This kind of metaphysical stances and their relation to the free will problem will be discussed in the next chapter.

Summary

As we could see in this chapter, the notions of freedom and moral responsibility *per se* do not create issues on the ground of Cartesian dualism. The problem as I see it is rather to sanction this metaphysical view – if somebody accepts Cartesian dualism in the first place, they should not have any problems with freedom and responsibility as they were defined by Descartes. It will be very visible later, when we will see critics of Descartes saying that free will as Descartes defined it is no longer possible (but not that it is an incoherent concept). The problems of sanctioning this view come up of course when we try to relate it to modern science, but it has also internal difficulties like the famous interaction between the mind and the body. All of these are driving factors of the stances that I will discuss in the next chapter – broadly called by me "materialistic" stances.

I see property dualism as a step towards these materialistic stances and therefore I briefly analysed its alleged advantages over Cartesian dualism. I do not see many of them both in regard to the notions of free will and moral responsibility, and the coherence with science. I see this kind of metaphysical stances as highly problematic stances because of trying to combine the best of two worlds – Cartesian dualism and materialism – but have big problems to deliver what they promise.

In general, the main problem of dualist stances in regard to freedom is how to make it compatible with the naturalistic view on nature. Cartesian dualism is highly biased towards preserving as much of humanity as it is needed to retain the Cartesian view on freedom. Let us now see the stances that are, on the other hand, biased towards naturalistic explanations of every possible phenomenon, including free will and morality. Afterwards we will also better understand the advantages and issues of dualism presented in this chapter

Chapter 3: Free will and materialism

Introduction

Here we arrive at, I believe, the core of this thesis. In this chapter I am going to examine important and interesting, from my perspective, materialist views on the relation between the mind and the body with respect to the notion of human autonomy.

There are two things I have to explain at the very beginning. First of them is why I consider these stances materialistic. In a few cases I will have to make a more detailed argument about it, but in general I believe all of them are physicalist in their core – physics is the ultimate science for all of them and everything that happens in the world is physical, even though high level descriptions may drastically differ. I call these stances materialistic, not physicalistic, mainly for convenience, since I want to avoid yet another debate over definitions and it is not uncommon to use these terms interchangeably (e.g. Stoljar (2015), Murphy (2013)). An important factor that makes them similar to one another is that they are in opposition to Cartesian dualism and from that perspective they seem to retain the notion of material substance and try to explain out existence of souls by ascribing functions of the soul to the matter. All of these stances are also naturalistic, but it is yet another term that is not easy to make sense of when comparing to materialism and physicalism. Whenever it is needed, I will tackle all ambiguities in relation to a particular stance.

Another thing is why I believe these stances are particularly important to the problem I am tackling. The answer is that they all provide a different kind of response to Cartesian dualism, and it gives a possibility to have a highly variable accounts of free will or human autonomy – we can see that these notions have indeed extremely various meanings and examine which of these meanings actually make sense. I will discuss various materialistic stances in the order from most "Cartesian" to the least "Cartesian" one. Or, in other words, as we will see, the stances I will present gradually diverge from Cartesian dualism so that in the end we get rid of not only the notion of soul, but also of important functions of the soul. As I will show in the next chapter, this process is, probably paradoxically, necessary to provide a sensible notion of free will I will try to defend. But before that, let us start with the first of the stances I want to discuss in this chapter. The first one is biological naturalism of John Searle.

Biological naturalism

It took me quite a bit of time to decide whether Searle's stance about the nature of the relation between mind and brain is dualistic or materialistic. It is not a surprise, because Searle considers it avoiding both dualism and materialism (Searle, 2004, 124), so assigning this stance to any of them is against the author's will. But conforming to an author's expectations is not a thing that we should be worried about when doing a philosophical analysis. Searle makes an extensive case against dualism and materialism, but in the end I believe his stance is materialistic. Let us start with describing what Searle wanted to say.

According to Searle the philosophical solution to the mind-body problem is very simple. Every mental state is caused by neurobiological processes in the brain and is realized in the brain (Searle, 2007a, 41). To be realized in the brain means in this case to be its higher level or system feature. It is sometimes referred to as supervenience of mental on physical, where supervenience is understood in the following way:

> A set of properties A supervenes upon another set B just in case no two things can differ with respect to A-properties without also differing with respect to their B-properties. (McLaughlin and Benett, 2018)

If we apply this notion of supervenience to brain and mind states we infer that any change in mental properties takes place due to a change in brain properties. And two brains that are in the same state will have the same mental properties. The "dualist" twist in Searle's materialism is that according to him we are not capable of examining states of mind from the outside – even though the difference in the states of minds stems from the difference in brain states, we cannot infer anything about the difference in states of mind from the difference in brain states. To explain this idea Searle uses notions of first-person and third-person ontologies. Third-person ontology is used to describe entities that can be perceived by an external observer, like trees, stones, human bodies, etc. As an example Searle gives water (2004, 98) – it is H_2O and there is nothing additional that water consists of and there is no difference between water and H_2O molecules. On the other hand, there is also a first-person ontology, which refers to entities that exist only insofar as they are experienced by some subject, human or animal. We would have no idea of conscious experiences if we were supposed to derive them from physical states, but the entities that are described using this first-person ontology are nevertheless a part of physical world. Speaking of two kinds of ontologies in one physical world resembles property dualism. What Searle gains by this terminological difference is that he may quite easily ascribe causal efficacy to mental states – they are capable of making a difference in the physical world because they are caused by physical states and are part of physical world – this is the very idea we touched upon in the previous chapter. Once again it is hard to explain where the difference between entities that are described using first-person ontology and third-person ontology comes from and why not to employ some identity theory instead[13]. Of course Searle has a reason to use this new terminology. He does not believe a mental state can be eliminated by reducing it to a physical state, because there is no doubt that consciousness exists and an eliminative reduction would imply that consciousness is just an appearance (Searle, 2004, 123). The problems that we encounter here seem to be typical of stances that try to combine best from dualism and physicalism. In such cases the relation between physical and mental and especially the ontological status of the mental is always suspicious – I will deliberate a bit more on that when discussing anomalous monism, the stance proposed by Donald Davidson. It is always hard to explain how exactly mental differs from physical or how physical "produces" mental. Here, Searle, instead of providing an answer to the question what it exactly means that mental states are causally dependent on physical states,

---

13 For example, Searle makes an analogy between consciousness and solidity as two different high-level features of physical entities – brain and wheel. He claims that consciousness is not ontologically reducible to physical microstructures, but not because it is something additional to physical microstructures, but because it has first-person, subjective, ontology. I do not see any reason to think this is a better and ontologically less expensive explanation than Cartesian Dualism which Searle refutes. Some argue that it is a form of dualism (e.g. Corcoran (2001)).

he delegates the question to neurobiology (Searle, 2007a, 40), so I do not see a reason to further investigate this specific issue – until neurobiology solves it for Searle, it is still an unresolved philosophical problem.

In the context of what Searle says about free will it is important to know the background of his theory, because of course his theory of mind influences what he says about free will (actually Searle emphasizes that what he calls "the problem of free will" is a problem about certain kind of human consciousness (Searle, 2007a, 45)), but also because he wants to do a similar thing to the "problem of free will" – he would like to delegate the problem to neurobiology. But let us first briefly talk about the idea of "free will" that Searle talks about. For him free will is a phenomenon that shows up in situations of decision making. At least sometimes when we have to make a decision, we experience that our reasons to make a particular decision are not causally sufficient conditions to make it happen. He calls this experience an experience of explanatory gap – a gap that occurs between fully determined segments in processes of deliberating, deciding and acting, and between each of these processes. For example there is a gap between a decision to do something and to continue it until the aim is achieved, as when somebody decides to learn a foreign language. Although the existence of the gap in question is very compelling, the same we can say about determinism, which in Searle's view is a denial of free will: "nature is a matter of events occurring according to causally sufficient conditions" (Searle, 2007a, 46).

Before Searle is able to answer the question "How is free will possible?" he has to answer another question: "How consciousness can move bodies?" In turn, he tries to explain the possible existence of free will by employing the notion of higher-level phenomena, one of which, according to him, is consciousness. He claims that our bodies are moved by consciousness, but under the hood these are neurons that make our bodies move – as consciousness is just a state they are in. The first-person ontology of consciousness is irreducible to neurons, but according to Searle, consciousness has no causal powers except the ones that it inherits from the neurobiological structures. This is why consciousness can act upon physical entities. As Searle says (2007a, 50), when he has a conscious intention to raise his arm, it is this intention that causes his arm to go up, but this intention is a feature of his brain system and therefore it is an action caused by neurons. So the question of free will turns out to be a question about explaining how neurons can possess free will when they are in some particular state called "consciousness," while retaining their purely physical constitution.

To be able to make a connection between the free will and an action that is performed by, after all, neurons, Searle employs an idea that reasons upon which we act are not ordinary causal explanations as in case of other physical events. In other words, they are not deterministic. As an example he considers differences between three explanations:

1. I punched a hole in the ballot paper because I wanted to vote for Bush.
2. I got a bad headache because I wanted to vote for Bush.
3. The glass fell to the floor and broke because I accidentally knocked it off the table.

Searle argues that the logical structure of the first one is different than the logical structure of the second and the third. In the sentences 2. and 3. the structure is "A caused B," whereas "because" in 1. does not imply that a desire to want to vote for Bush forced somebody to punch a hole in a ballot paper. One could have the same desire to vote for Bush and do not punch a hole in a ballot paper. Sentences 2. and 3. contain causally sufficient conditions for something to happen, while there are no such causally sufficient conditions in case of 1.

If we accept such a view, then a new problem arises – rational explanations seem to not explain anything, because when somebody says "I did A because of B" there must have been a possibility for ~A to happen even if B occurred. Searle employs the idea of self to explain how it is possible. He claims that a sentence of a form "Agent S performed an act A because of reason R" is of the form "a self S performed action A, and in the performance of A, S acted on reason R" (2007a, 53), which is supposed to be quite different than the "ordinary" causal explanation. Searle's conclusion is that rational explanations of actions require us to postulate the existence of self that is irreducible to its neural background. What is more interesting, he thinks that by adding two assumptions we can derive (not only postulate) existence of this irreducible self. These two assumptions are: (1) "Explanations in terms of reasons do not typically cite causally sufficient conditions." and (2) "Such explanations can be adequate explanations of actions" (Searle, 2007a, 53). Assumption 1 is extracted from what Searle describes as acting on reasons. He claims to know that the assumption 2 is true because he knows from his own experience that sometimes he acts upon reasons and those reasons are adequate explanations of his actions. To make his point he adds Assumption 3, which reads: "Adequate causal explanations cite conditions that, relative to the context, are causally sufficient" (2007a, 53). From Assumptions 1 and 3 he derives that reasoned explanations are inadequate if we consider them as ordinary causal explanations (Conclusion 1) and therefore they must be considered as not ordinary causal explanations (Conclusion 2). And these non-ordinary causal explanations are nevertheless adequate, because they explain why a self-acted in a certain way when it experienced a gap between reasons and actions that Searle claims is necessary for a free action. As he describes it, it cannot be a Humean self, because a bundle of perceptions would not be enough to account for the adequacy of explanations which is the reason why a self has to be taken into account. As Searle notices, the conclusion that there is such an irreducible self does not follow from the assumptions, he claims to make a transcendental argument: the existence of irreducible self-capable of acting on reasons is the condition of possibility of adequacy of rational explanations. Searle notices one problem with his notion of adequate explanations: we can always ask why something was an explanation of an act. E.g. if somebody asks, "Why did you vote for Bush?" one may answer "Because I wanted an improvement in educational system." and it may be followed by another question, "Why did you want an improvement in educational system?," and an answer to that question may be followed by another one and so on. According to Searle the fact that an answer may beg a new question does not mean that it is inadequate, because "explanations (…) have to stop somewhere" (Searle, 2007a, 56).

The next step for Searle is to provide an idea of how free will can be realized in human brain. In other words, how the relation between free will and human brain can be similar to the relation between consciousness and human brain. In the picture outlined above it is assumed that there are gaps between decisions and actions, but in the lower level of decisions there are only firing neurons which cause other neurons to fire and there is no gap that would resemble what we experience in the higher level. According to Searle, when we deliberate and make a decision there are two possibilities: either the lower-level brain states in which the deliberation process is realized are sufficient to cause the brain states in which action is realized or not. If they are sufficient, then we have no free will and if they are not, we do. In the first case free will caused by the explanatory gap is just an illusion.

Let us get deeper into the second case. If we have two brain states at two different times t_1 and t_2, the state of the brain at t_1 is not causally sufficient to determine the state of the brain at t_2. Instead, the change from the state at t_1 to the state at t_2 is mediated by consciousness, which is at the same time determined by the state of neurons in which it is realized. To make it clearer, the state of neurons determine the state of consciousness, but this state of neurons itself is not enough to explain the transition from the state at t_1 to the state at t_2. To do that we need to employ conscious deliberation that happens on the level of consciousness, and it is the reason why the change from the state at t_1 to the state at t_2 happens. But it is of highest importance for the Searle's project to remember that the state of consciousness at any point in time is entirely fixed by the behaviour of neurons, because otherwise it would not be possible to overcome Cartesian categories of distinct immaterial soul and material brain that for Searle are the reason of stagnation in philosophy of mind.

Based on his deliberations summarized above, Searle gives three conditions for the brain to function in accord with a possibility of free will: (1) Consciousness functions causally in the moving body. (2) The brain causes and sustains the existence of a conscious self that makes rational decisions and applies them in action. (3) The decisions and actions of the conscious self cannot be determined in advance by causally sufficient conditions, but still they are rationally explained by the reasons of the agent on which he/she acts. The main problem is of course with the third condition, because, according to Searle, it involves adding "rational indeterminism" to how brain functions. He believes the only way it could be done is by employing quantum indeterminism, because all indeterminism in nature we know is quantum indeterminism.

In terms of free will, Searle seems to be a philosopher who gets at least one thing right – he says it explicitly that the free will problem needs many other philosophical problems to be solved e.g. the nature of consciousness, causation or scientific explanation and rationality (2004, 215). He also acknowledges the seriousness of the problem, unlike other philosophers that end up with a "theory" of mind, but do not touch the problem of free will that this theory implies (like for example Chalmers). But it is not hard to see why his "solution" is not really convincing. I believe the considerations above show how much confusion is caused when somebody tries to reconcile naturalism with existence of rational selves making free decisions – there are a lot of inconsistencies and suspicious steps that could be easily replaced by something more plausible. Let us start with the idea that there is an explanatory gap between our decisions and our actions (or our deliberations and our decisions). As an example of such an explanatory gap Searle gives a situation when somebody wants to vote for Bush and punched a hole in the ballot paper because of that. The reason for making a claim that there is an explanatory gap involved here seems to be simple: it is possible that somebody else wants to vote for Bush and does not punch a hole in the ballot paper. But it seems to be superficial. If, as Searle claims, consciousness is just a particular state in which neurons are and a conscious self, also being a result of brain processes, is what makes a decision to punch a hole in the ballot paper, it seems much more plausible that the consciousness of a person that punches a hole in the ballot paper and the consciousness of a person that does not do that are embedded in brains with different properties and this is what causes difference in their actions[14]. They both have (at least) one similar property – they want to vote for Bush. But the brain is a system that works as a whole, and we cannot isolate one of its processes from others. What is startling, Searle is aware that we may deceive ourselves in different situations – for example one may claim to make a rational decision which is in fact influenced by emotions, prejudices, etc. – but he says that in ideal conditions reasons are adequate explanations of actions. The problem is that probably nobody is ever in such ideal conditions, there are always some unconscious processes in our brain that influence our decisions – only a self that is not embedded could be free in this view.

---

14 This explanation can be easily achieved using eliminative materialism that is discussed later in this thesis.

Above considerations lead to the conclusion that maybe Searle should say that reasons explanations are actually not adequate. One could say that it is better to talk about two kinds of adequacy – one that involves a conscious self that acts on some particular reason and another that requires a full explanation of an action. But it seems that the first one is just a veiled version of the second one – reason explanations are adequate because under the hood they are ordinary causal explanations, so by themselves they are not adequate. Another problem that shows up is the suspicious self. It is hard to describe what it actually is and how it is somehow independent from the brain (as ontologically not reducible to it) while being embedded in it. Searle claims that it cannot be a Humean self, but if all these critical remarks are correct, Humean self is perfectly suitable. Humean self is just a bundle of perceptions, and we have to remember that it is a bundle of perceptions in some particular time. In other words, we can think of different bundles of perceptions as different selves. Applying this remark makes it easy to explain how it is possible that not only different people make different decisions in seemingly the same situations, but also why a particular person makes different decision in the same situations in different time.

There are of course other problems with this stance. For example, there is no reason to think that the explanatory gap we perceive in the first-person ontology view is the same as one that exists on the neurological level, if it indeed exists. Searle's view is some strange version of naturalistic agent causation (as he refers to self as the reason of making a particular decision), but due to the fact it is naturalistic, and it relies completely on the premise that self is ultimately a result of how brain functions it is not convincing to use it in an explanation of free will. Also saying that we do not have to worry about possible further explanations of our choices is handwavy. To make this theory work, Searle would need to employ some more straightforward version of dualism. It is possible to interpret his stance in dualistic spirit, but then all the merits it is supposed to bring would be probably lost. But even Searle makes it clear why his stance has more of materialism than dualism in it, when he asks a question that is supposed to point us to the main problem of the division of reality into mental and physical states: "How do qualitative, subjective, and intentional phenomena fit into the physical world?" (Searle, 2004, 128). If the question was posed in the other direction (e.g. "How do quantitative, objective, and non-intentional phenomena fit into the mental world?"), it would point at a spiritualistic solution. To be fair, Searle is concerned with the fact that physical phenomena that we talk about in contemporary physics could be non-physical in terms of what Descartes claimed to be physical – "if electrons are points of mass/energy, they are not physical on Descartes' definition because they are not extended" (2004, 128) – but there is no reason to think that Descartes would oppose contemporary physics and the shift in the way we think about physical entities or call some physical entities "mental".

Speaking about Descartes also one more important thing needs to be mentioned. It is not entirely clear that the way Searle speaks about making a decision based on reasons is a sensible way to express the notion of free will. It can be thought of as the lowest degree of freedom mentioned by Descartes and at best it is a very weak notion of free will. But since Searle's stance has other theoretical problems, there is no need to explore it further.

All these critical remarks could be summarized by one simple sentence: adding a level of indirection does not change the essence of the whole process of decision making. In computer science there is a wording: "We can solve any problem by introducing an extra level of indirection." But in computer science nobody believes they introduce something qualitatively new, adding an extra level of indirection let us make easier operations on abstractions, but everything works in qualitatively the same as if the operations were done on a lower level. If anything, by adding an extra level of indirection we can rather have a drop in performance. This level of indirection in Searle's theory is of course the self which is a feature of the world caused by the brain. Except for the suspicious idea that something may be causally but not ontologically reducible to physics, "self" is just a name for brain processes and there is no reason to believe that it may have any kind of properties that could account for free will as Searle understands it. If the brain is a fully determined system, then deliberations of a conscious self, which is a brain in some particular state, must also be fully determined. On the other hand, if there is indeterminism at play, it is still the case that the existence of self is caused by the brain processes, there is just some randomness involved, which also cannot account for free will, even when it is defined in the way Searle does it.

Anomalous monism

Instead of making an abstraction over brain states and assigning a label "self" to it, another option is to say that brain states and physical states are identical. This is the path chosen by Donald Davidson in his formulation of his own stance with regard to the psychophysical problem, anomalous monism. I believe it is instructive to describe his view on the relation between the mental and the physical as the next step, as it is a stance somewhat in between dualism and biological naturalism on one hand and more reductive stances like eliminative materialism (described later) on the other.

Davidson describes his anomalous monism as materialism without assuming that mental phenomena can be given purely physical explanation, which, according to him, is a standard assumption of materialism (Davidson, 2001, 214). On the ontological level anomalous monism is an expression of a belief that all mental events are identical with physical events, whereas converse is not true. In other words it allows a possibility that not all events are mental, whereas all events are physical. One of the outcomes of his theory is supposed to be a possibility to deny that claims typical of reductionism, claims whose general form can be expressed as "x is just nothing but a complex neural event", where x is some mental event (an example given by Davidson is "Conceiving the *Art of the Fugue* was nothing but a complex neural event" (Davidson, 2001, 214)). Regarding the free will, anomalous monism is supposed to introduce a concept of anomalousness of the mental, hence the name. The aim of this subchapter is to examine whether anomalous monism delivers what Davidson promises.

Anomalous monism is a stance that is supposed to reconcile three principles that are commonly perceived as contradictory. The first of them states that at least some mental events interact causally with physical events. "At least some" is just to account for a possibility of mental events that are neither caused by, nor cause physical events and if such mental events exist, they are irrelevant to the argument. The second principle asserts that wherever there is causality, there must be also a strict, deterministic law that binds the cause and the effect. From the assertion of the first and the second principle it seems that we can already derive that the relation between mental events and physical events that interact causally must fall under a strict deterministic law. But here comes the third principle, which is that there are no deterministic laws on basis of which mental events can be predicted and explained (Davidson, 2001, 208). This is also what the anomalous character of the mental events is. The "alleged" contradiction is of course between the first two principles and the third one.

How can these three principles be reconciled? To do that Davidson employs the difference between an event itself and its description. What we call a "mental events," or a "physical event" is actually the same event having two different descriptions – mental and physical. Both of them use different and, what is important, incompatible with each other language. It is incompatible in a way that it is impossible to derive one description when we know the other – which is just a restating of the third of the principles I mentioned. So a mental event is identical to a physical event and this physical event is bound by a strict law to other physical events – ones that are its cause and effects. What is important for this theory of identity is that identical can only be individual events – if we could speak of identity between kinds of mental and physical events, then it would mean we are able to construct laws that bind them. So they are identical in a way that every mental event can be singled out using a physical description.

As an example of why it is reasonable to speak about anomaly Davidson gives the failure of definitional behaviourism. It is impossible to give definitions of mental events in terms of behavioural ones, because the meaning of behaviour has to be vindicated by qualification using mental terms that will make sure that the translation is accurate. Following the Davidson's example, if somebody answers "Yes" in response to the question "Is there life on Mars?," it does not have to mean that they believe there is life on Mars. They need to understand English, their response has to be intentional, it has to be a response to that very question, they need to understand the question in the same way the asking person does, etc. To put it in other words, the exact same behavioural reaction can be related to different mental states on different occasions – we can always imagine a counterexample of somebody in different mental state, but the same physical state. Of course, we can further patch each of these responses with behavioural conditions to make sure the response is intentional, etc., but it is a process that can go to infinity, at least according to Davidson. He claims that even if we found a behavioural description that is exactly coextensive with some mental term, nothing would be able to persuade us that it is so, as what we know (probably from our personal experience, not scientific investigation) makes it hard to us to trust any general statements linking behaviour and thoughts.

Above remark shows what is a bit similar in anomalous monism to the Searle's condition of rational explanations – rational explanations are reasons to act upon for some particular agent and it is also impossible to get the ultimate response why an agent acted upon some reason. What is however very different, is that Davidson does not refer to a self as a source of mental descriptions (whatever it would mean), although, as we will see later, he makes a brief remark on a concept of a person his theory helps to make sense of. What is the distinguishing feature of the mental is not anything that can be ascribed to a self (at least not directly), like being private, subjective or immaterial (examples given by Davidson), but exhibiting intentionality, which is here understood as an ability of a mental state to refer to something (Davidson, 2001, 211). In a way, what Davidson does is also similar to a transcendental argument. To be able to treat others as persons we need to find coherent and plausible patterns in their attitudes and actions (2001, 221–221).

Now, turning to the problem of free will, how does all of this help Davidson to retain a sensible notion of free will? Probably, all we can say is that even though every mental event is identical with a physical event, we cannot make two kinds of predictions. First, we cannot predict mental events based on our knowledge of physical events, because there is no lawlike connection between them. Second, we cannot predict mental events based on other mental events. I think this language is very vague, as there is no dualism of events and predicting mental events based on knowledge of physical events implies some sort of dualism, at least in the language. Let me restate this theory using different notions. There are events. All events have physical description, some also a mental description. All events are bound by laws, no exceptions. But all laws are stated in a language and in this case it is a language of physics. So we can, at least in principle, predict every event, but only if we use a physical description to refer to it. What we cannot do, on the other hand, is to describe this event using terms of mental language based on the description of the same event in the physical language. Also, when we have a mental description of an event, we are not able to derive the physical description associated with that event. This is the anomaly that Davidson speaks about but expressed without using quasi-dualistic language.

How is that helpful in stating that human beings are free? I see this as freedom of language incongruence. We cannot make translations from mental language to physical language and the other way around. But we get to the same problem as we got when considering Searle's biological naturalism. Every event is possible to be predicted using strict deterministic laws. Knowledge of mental language does not give us any advantage in predicting events, because they all can be predicted using physical theories and physical language. Mental events are causally effective in the physical world, because they are physical events. Of course, it solves the problem of causal inefficacy of mental events and the anomaly is supposed to solve the problem of dependence of the mental events on the physical events. But by turning from mental and physical substances to mental and physical descriptions, which is basically another way of escaping Cartesianism, the very essence of that distinction seems to be lost.

Davidson's argument entails that without mental descriptions of events, we fail in treating others as persons. But maybe there are no persons – only physical objects involved in physical events? Does it not mean that the concept of a person is not needed in our description of the world? It is for sure not needed in the physical description, but non-reductionist physicalists like Searle or Davidson will insist that since it is impossible to reduce first-person ontology to third-person ontology or mental descriptions to physical descriptions, respectively, and because of the supervenience of the mental on the physical we can retain both a meaningful concept of a self or a person and self's or person's causal efficacy.

There are two ways a reductive physicalist can respond. First is to show that non-reductive physicalism fails at showing that mental states have causal efficacy. Second is to present a stance that succeeds at reducing concepts of a self and person to the physical language.

The first path is taken for example by Kim in his famous Supervenience Argument (Kim, 2008, 39–45). This argument's purpose is to show that the assumption of supervenience makes it impossible for mental states to be causes of physical states. In short, when a mental state m_1 supervenes on a physical state p_1 and is supposedly a cause of a mental state m_2, then it is actually p_1 that is the cause of m_2, by causing p_2, which m_2 supervenes on. If this were not true, then there would be two principles that would be violated. First, the causal closure principle we discussed earlier, second, the exclusion principle. The exclusion principle states that no single event can have more than one sufficient cause occurring at any given time, unless it is a case of causal overdetermination. Kim of course assumes p_2 is not overdetermined and that is why this principle would be violated.

I will not get into the details of this argument, as it is based on the typical reformulation of the terms involved in the psychophysical problem we have already seen so far. In my opinion, Kim also does not manage to establish identity of the mental and the physical states – Davidson could argue that m_1 causes m_2 because m_1 is p_1. What seems sure to me is that it is impossible to retain physicalism and a sensible notion of mental states being not reducible to physical states. Therefore I will focus on presenting a solution to the psychophysical problem that gets rid of all mental terms and states all at once – eliminative materialism.

Eliminative materialism

The thesis of eliminative materialism was very clearly expressed by its author, Paul Churchland, in his article *Eliminative materialism and the propositional attitudes*:

> Eliminative materialism is the thesis that our common-sense conception of psychological phenomena constitutes a radically false theory, a theory so fundamentally defective that both the principles and the ontology of that theory will eventually be displaced, rather than smoothly reduced, by completed neuroscience. (1981, 67)

This short passage is of course very general, but it is extremely packed with meaning. We can already see that if the thesis of eliminative materialism is right, then both Searle's idea of free will based on irreducible self and the idea of autonomy based on anomaly defended by Davidson does not make much sense. I believe this is the only possible coherent physicalist explanation of the psychophysical problem, because it gets rid of the mind entirely and reverts to the brain as the ultimate explanation of human actions. The aim of this subchapter is to show how it is achieved and what are the consequences of that.

There are two things which I want to discuss regarding the first part – the disposal of mind. First is what Churchland calls "folk psychology" – presumed theory of mind whose ontology Churchland wants to get rid of. Second is what this ontology should be replaced with and how. To do that I will refer to recent inventions in artificial intelligence on which eliminative materialism is built.

Since eliminative materialism stems from critique of folk psychology, it is inevitable to start considerations about this stance with a comparison regarding folk psychology and eliminative materialism. But to be able to do that, the very notion of folk psychology needs a little clarification. Churchland claims it is a theory and as a theory it is supposed to help us with organizing the reality under laws binding entities that to this theory belong. According to Churchland the areas it is supposed to help us understand are the major problems in the philosophy of mind, like prediction of behaviour, existence of other minds, intentionality of mental states or the mind-body problem (1981, 68). Folk psychology is supposed to achieve these aims using ontology whose fundamental part are propositional attitudes. The very term "propositional attitudes" was coined by Russell in his work *The Philosophy of Logical Atomism* (2010, 60), although he prefers to talk about "propositional verbs," because it does not presuppose reference to psychology and some verbs that have a form of relating an object to a proposition are not (in this way) psychological verbs. But the key point is the form of relating an object to a proposition. As Churchland notices, on the ground of folk psychology this relation is not "a mystery of nature" (Churchland, 1981, 70), but rather an obvious fact. Mental states are "by definition" intentional, which means that they are "about something," to refer to a widespread definition of intentionality.

I have already referred to the same idea while describing what, according to Davidson, distinguishes mental and physical events. Let us now discuss this idea in a bit more depth. Propositional attitudes discussed by Churchland relate an object to a proposition. For example, in a sentence "Margaret thinks that John is the worst student at the university." the verb "thinks" binds Margaret to the sentence "John is the worst student at the university." It also expresses the nature of this binding – we could replace "thinks" with "hates," "likes," "doubts," etc., and there would still be binding between Margaret and the same sentence, but its nature would be entirely different. One of the peculiar characteristics of propositional attitudes that makes them seemingly inevitable in our language is a logical property that when we substitute a term in the embedded sentence with another term with the same extension, the truth value of the whole sentence may change. For example if in the above sentence I change "John" with "the captain of the university football team," the sentence "Margaret believes that the captain of the university football team." may be false if Margaret does not know that John is the captain of the university football team. It is called by Churchland an anomalous logical property (Churchland, 1981, 70) and he believes some philosophers were inspired by it to believe that some parts of human activity cannot be explained referring only to physics. According to him, it is important to treat folk psychology as a scientific theory, because if we do so, intentionality is not some kind of a mystery of nature which has to be further explained, but an indispensable part of the theory.

Here we come to the three possibilities that Churchland considers in relation to what may happen to the propositional attitudes when the neuroscience will be completed. Intentionality may be either reduced to completed neuroscience, proven irreducible to completed neuroscience, but preserving its indispensable status and completely extinguished from our scientific theories. Of course, eliminative materialism takes the last path – intentionality is supposed to be neither indispensable, nor reducible through inter-theoretical identities, hence it is eliminated.

There are two things that are especially important when it comes to this elimination. First is of course how it happens or what intentionality is replaced with. Second is what the advantages of elimination are with respect to folk psychology.

Let us start with the first thing. Intentionality is eliminated because the brain is supposed to implement structures that cannot be representations of intentional states. It is basically all there is to it, but of course we need a more detailed description of what is supposed to happen in the brain. Brain is supposed to implement neural nets. Neural nets are just very complicated mathematical functions and as any function they take input and produce output. In case of a neural net implemented in a human brain an input can be a stimulation of a sense organ and an output is some concrete behaviour given that stimulation. The stance that the brain implements a neural net is called connectionism and to understand how the neural nets can be a threat to intentionality we have to discuss different variants of this stance.

But talking about what neural nets do is like dancing about architecture – to really understand what happens under the hood and why it is supposed to extinguish the notion of intentionality it is necessary to see a neural net, at least very simple, in action. The neural nets that are inspiration for eliminative materialists, or more generally connectionists, are artificial neural networks implemented in computer's hardware. It is not important what kind of a hardware it is – it can be a CPU with a separate RAM or a GPU with its own RAM – the character of computations is the same. What they do is take a tensor representation of an input, transform it multiple times using various tensor operations and in the end produce an output based on those transformations. What is really the most important is what happens in the middle – transformations let us project input tensor onto space with a different number of dimensions and after various transformations applied to the input sequentially it is possible to produce the output – for example classify an image. Let us take a look at what a simple neural net classifier does. Let us say we have one of the digits from the MNIST dataset, a dataset of handwritten digits, to classify. This digit can look like this:

A human being should not have a problem with classifying this digit, it is clearly 3. But to a machine it is provided as a 28x28 array of integers. In the upper left corner for example all integers will be 0, because that area is purely black. Closer to the middle of the image the array will have values closer to 255, because that part of the image is white. So a neural net has as its input 784 values (28 * 28) and has to return one as an output. But even outputting one value is not that simple either – a neural net trained on the MNIST dataset will typically output 10 values – probabilities that a particular digit is 0, 1, 2, etc. A properly trained classifier will output high probability for the digit 3 and low probabilities for all other digits. So in the end, that is all what such a neural net can do – take as an input an array of 784 values and return an array of 10 values by using mathematical transformations in between.

One thing that needs to be explained before discussing how this all is supposed to be a threat to intentionality is what the training means. I think the word "training" is actually a bit obscure. What happens during training is fitting the function that is expressed by a neural net to the examples that are shown to the net. And since the function is expressed using tensors with numbers that are used to do transformations, fitting means changing these numbers so that the final output fits better the expected outcome – in our examples the number that is represented in a picture. These numbers are called weights.

For the argument I want to make it is not necessary to know all the details of how neural nets work, are trained, etc. The most important is to understand what happens in an already trained net – after all we are interested in what happens in a brain of an adult person when they are about to make a decision. The network from the example above outputs a digit from 0 to 9 given a picture with a handwritten digit. At this stage there is nothing indeterminate that happens in the network. The network deterministically produces the output given an input, even if we are not able to predict what the network will output ahead of time – to be able to do that we would have to do all of the computations ourselves.

After these preliminary remarks let us think what it means that the brain implements neural networks for language understanding. This is an idea presented for example by Ramsey, Stich and Garon in their paper *Connectionism, Eliminativism and The Future of Folk Psychology* (1990) where they are trying to show that if a certain version of neural nets is the correct representation of what happens in human brain, then eliminativism in regard to propositional attitudes is a correct stance. I am going to present both their stance regarding the above conditional and counterarguments to it, but at the moment what I want to focus on is presenting the very idea of a neural net representation of language that they use.

Ramsey, Stich and Garon trained two neural nets with the same architecture to determine whether a sentence is true or false – why two I will get back to in a moment. The sentences they used had a form X have Y, where X is a word referring to an animal species and Y is a feature that animals of this species have, for example "Dogs have fur." or "Cats have gills." After training, both of the neural nets could perfectly classify sentences as true or false. The point of this modelling was to show that neural nets can accomplish this task without functionally discrete, semantically interpretable states which can be interpreted as representations of propositional attitudes that play a causal role in production of other propositional attitudes and, subsequently, a behaviour. That is the reason why two nets were trained – one was trained on 16 sentences and the other on the same 16 sentences plus additional one. The result of this were two neural nets that can perform well on the same 16 sentences but having different weights and biases due to the one sentence present during the training of one and absent during the training of the other net.

According to the authors, these models share three properties: information is distributed among weights and biases, the connections between weights and biases do not have symbolic interpretation and they are supposed to be cognitive models, not just implementations of them. This description is of course aimed against symbolic information processing. In given example the same sentences could be represented symbolically, where the truth value associated with each sentence would be computed based on explicit determination of presence of distinct words in a sentence. In case of neural nets information is distributed because we cannot determine the truth value of any sentence based only on selected nodes – it is encoded in a whole net. Information is also subsymbolic, because it is impossible to determine in which part of the net each part of a sentence is encoded and what are relations between those parts. The third property to which authors refer is not distinctive to all neural networks – to say the models they used in the article are supposed to be cognitive models, not mere implementations is just to stress that it is possible to implement symbolic models using neural nets and it is not what the authors are talking about in the article.

This last remark is helpful in determining why the model of cognition represented by the models in question is supposed to be dangerous for propositional attitudes. In case of symbolic representations (both using neural nets and in other implementations) it is possible to determine which conviction was causally important in producing other convictions and behaviour. For example if we want to determine the truth value of a sentence "Dogs have fur." we can easily determine how presence of each token contributes to the output "true." On the other hand, if we pass representation of this sentence through a neural net, we can only get an answer for the whole sentence, and it does not make sense to look for relations between the words. This applies of course also to more complicated sentences and production propositional attitudes. The expected effect of this is to be able to show that it makes no sense to ask whether a representation of a particular proposition plays causal role in the output of a network. It is not even possible to recognize a representation of a particular proposition in a network. It is clearly visible when we compare weights of the two networks I mentioned before. As Ramsey, Stich and Garon point out, these networks have no projectable features in common (1990, 514) and there is an infinite number of neural nets that could represent the same sentences. They conclude that networks that can model cognitive agents with capacities to believe the sentences are true/false are not a genuine kind, but rather they belong to a chaotically disjunctive set (1990, 515), whereas folk psychology treats all people with conviction that, for example, dogs have fur as having something in common.

There are two objections that authors discuss, and it is advisable to at least briefly discuss them here. The first is that the discussed models cannot be seriously treated as models for human belief or propositional memory. Opponents say they do not generalize well and will not scale up to many more sentences. I do not wish to discuss this objection in great detail due to two reasons. First is that the contemporary advancements in neural networks help in being confident that we can build a neural net architecture that is capable of solving any cognitive task that humans are capable of solving – whether they do it in the same way as humans do is, however, a completely different story. Second is that I am just simply going to assume eliminative materialism as the correct stance regarding the psychophysical problem and draw conclusions based on this assumption.

The second objection is that it is not true that the connectionist models discussed violate propositional modularity – the propositions are encoded in a not entirely obvious modular way. This is interesting, because it would render eliminative materialism devoid of meaning – even the authors admit that if this objection is true, then their argument would be seriously undermined (1990, 517). This is the path taken by for example by Skokowski in his article *Networks with attitudes* (2009) where he tries to defend a view that there are neural correlates of a belief. He argues that whereas it is true that the weights of a neural net do not represent propositional contents, there is a part of a neural net that does it in a straightforward way – these are the input units of a network. In case of humans these are the sense organs and the way they are wired into the brain (2009, 464). To make his point he compares what happens in a brain of an infant and a brain of an adult when their sensory organs are stimulated with a view of a flower. Sensory delivery systems of both an infant and an adult carry the same information, but it is the knowledge acquired with learning that makes different parts of the content available for an agent and therefore participate in computation of behavioural output. An adult that believes an object is a flower can do things that are possible to do with flowers.

Since beliefs have causal role in behavioural output production, they cannot be eliminated – this seems to be the expected result of Skokowski's argument. He gives another example – if he runs and sees a tree, he will swerve. He has a belief that there is a tree in front of him when he perceives it in appropriate circumstances, this belief causes his swerve and then it ceases to exist. This belief not only does not endure, but it also cannot endure, otherwise he would be swerving all the time (Skokowski, 2009, 466).

I do not think it is advisable to discuss this objection further – it is enough to just show what are the problems that seem to be already apparent. First of all, Skokowski does not show an answer to one of the replies we can find in the original paper by Ramsey, Stich and Garon, where they claim that the architecture of neural nets they discuss is not capable of expressing beliefs that are not currently active – somebody does not have to consider a sentence "Kangaroos are marsupials." to believe that that kangaroos are marsupials (1990, 518). It could be possible to claim that this reply stems from our wrong understanding of what beliefs are and we would be stuck with a question "is it reduction or elimination?," but in the end it does not look like something worth fighting for. After all, it does not change anything at all in our understanding of human behaviour and looks like an ad-hoc idea to force preservation of propositional attitudes. We can "see" propositional attitudes in brains just because we use them on daily basis, but except of the usage of the term "belief" what Skokowski proposes does not make any difference.

After these lengthy preliminary remarks regarding eliminative materialism it is time to get to the main reason to discuss this stance, of course implications for possible understanding of free will if any is possible. Here I will mainly discuss the work of Patricia Churchland in the area of something that she would probably call naturalized morality. Declaratively, she is an eliminative materialist just like Paul Churchland and therefore it makes sense to see how we can draw a connection between theoretical philosophy that is supposed to lead to eliminative materialism and naturalized practical philosophy.

Let us start with the notion of free will that she uses, or I should rather say the notion she wants to replace free will with. According to her we learn what the notion of free will means from experience, like every other notion (Churchland, 2006, 43). It has a clear reference to connectionism, where neural nets are supposed to learn new concepts by seeing examples of them. As with every other concept we see clear prototypical examples, like Chamberlain's choice to appease Hitler or a dreaming man who strangles his wife, which are free and unfree, respectively, but there are also examples that are ambiguous, like a mother that drowned her children in a bathtub, but then called a police because she understood her actions were against the law.

But it is not the main problem with the notion of free will for Churchland. She says: "A rigid philosophical tradition claims that no choice is free unless it is uncaused" (2006, 43) to then conclude that it is impossible to meet this condition, since decisions are made by brains and they go from one state to another in a fully determined fashion, based on their antecedent conditions. So she suggests that instead of talking about free will it is better to talk about neurobiology of self-control. It is supposed to also help us with ambiguous examples like the woman that drowned her children. As examples of self-control she gives a dog that learnt through reinforcement to lie quietly when the local squirrel taps the screen door for peanuts and a hungry chimpanzee that does not reach for bananas when the alpha male sees it. She does not define self-control, presumably because we can learn the meaning of this concept from examples, but it seems that self-control is an ability to do something different than one would like to do due to some reason. It is congruent with what she says about self-control in her book *Conscience*, for example:

> Self-control, dependent on regions of the frontal cortex, is crucial in inhibiting suboptimal choices, such as choosing immediate gratification, with the result that you forgo better, long-term rewards. Roughly, the more neurons in the frontal regions, the greater the capacity to control impulses. Even rodents, however, with their rather modest prefrontal cortex, can show impressive self-control.
> (2019, 85)

So instead of free will we end up with self-control and the main difference between them is that the second one can be (and is) completely determined. There is much more to say about the relation between these two concepts (and whether actually the concept of self-control is coherent and fits eliminative materialism), but I will discuss it a bit later. Now I would like to discuss what kind of moral responsibility Churchland wants to achieve with her new notion of self-control. According to Churchland morality is an instinct that we have gained via evolutionary processes with combination of self-control gained via reinforcement. She does not give any specific definition of morality, but everything she writes in her two books, *Braintrust* (2012) and *Conscience* (2019), leads to such conclusion.

What does it mean that morality is an instinct gained via evolutionary processes? The first thing is that we are social animals, and it is because the brains of our ancestors were adapted for sociality (Churchland, 2019, 25). Technically, according to Churchland, due to the process of evolution feelings of pleasure and pain supporting survival of oneself started to motivate affiliative behaviour and mammals started to care about others. Since their brains were "adapted" they must have gotten evolutionary advantages due to this process. There is also a question what "others" are the ones that mammals started to care about.

It is not a big surprise that the first of these others are babies and the mammals that got attached to them are those babies' mothers. Without this, human offspring could not survive so it is only logical that protection of one's offspring got somehow wired into human brains. According to Churchland there is even more to this – humans benefit from the fact that they are born helpless, and their mothers have to look after them. When humans are born, the neural nets must be as minimal as possible to maintain their life outside the womb, so that experience has bigger impact on their subsequent behaviour (Churchland, 2019, 31). This is the key to adaptive capacities of every human being as opposed to evolutionary adaptive capacities of other species. Other species need to wait for generations to adapt to new environment, while humans can do it multiple times within a lifetime. Because human brains are relatively big and they need to grow quickly after the birth, calorie intake of a human baby is huge and it is the mother, with help of the child's father, provides food. So basically what we call "morality" is supposed to be primarily a result of high calorie needs of human babies and lack of capability to provide for them by oneself. Afterwards it got spread on other people as caring about others and cooperation can help to achieve the ultimate goal of everybody's existence – their own genes propagation (Churchland, 2012, 32). She does not make this conclusion, but probably the end result of this whole process is our societies in which we live together – we agree to abide to some rules that are inconvenient to us when we live in a group because unconsciously our brains compute this behaviour as an output to the question "what behaviour will lead to the greatest possibility of my own reproduction?"

But this is just the first part of the story that Churchland tries to tell us. The second part is how we get to self-control that makes us follow the rules and how exactly are the rules "chosen" or selected. This is where reinforcement come into play. At this point we are already beings that care not only about themselves but also about others and therefore it is easy to trade a part of our own freedom for well-being of other people. We start to control our behaviour based on responses we get from others in our environment. When they react in a positive way we tend to continue or repeat our behaviour, while when they react negatively we tend to cease what we do and not repeat it. There is however one prerequisite to that – one has to feel shame or guilt to adopt their behaviour according to responses of others. Without it a person not only does not change their behaviour, but in the end also does not see there is anything wrong in what they do (Churchland, 2019, 129).

This is really almost everything there is to be about Churchland's stance regarding morality. In *Braintrust* and *Conscience* she talks much more about how exactly are the discussed processes implemented in human brain which is completely uninteresting from philosophical perspective, so I will leave these parts – after all, it does not matter whether bonding between a mother and a child happens due to oxytocin, vasopressin or any other hormone and what their number is, since the process is exactly the same. But there is one important thing to add, and it is the relation between Churchland's stance and what she calls "other approaches" to morality. On one hand Churchland claims that science is not capable of telling us what we should do, but on the other she strongly rejects other stances regarding morality, like for example Kantian or utilitarian. One could say that naturalized morality endorsed by Churchland refers only to the origin of moral feelings and there is a completely different area of human thought that embraces the realm of obligation and duty. This could be supported by the fact that Churchland says science cannot help us in determining what we should do, but the rejection of other stances makes it ambiguous. Also, naturalistic account of morality would be totally uninteresting from philosophical perspective if it were not to involve rejecting other stances in pursuing right normative claims. The most interesting, and controversial, part of this naturalistic account of morality is that it is supposed to explain where our moral feelings come from while maintaining that there is *no* source of right normative claims (like e.g. reason) and there cannot be one.

Where do our moral rules on which we base our law come from and how can we justify the fact that we judge others? Churchland says that we must hold individuals responsible for their actions, because we value our social life and there must be limits to maintain it (Churchland, 2006, 44). Here I will finish the summary of Churchland's view and get to the more important part – it's critique. At first I will make a critical analysis of this approach to morality as it is by itself and at the end I will refer to what eliminative materialism implies about it.

Let us start with the justification of moral responsibility suggested by Churchland. It is quite easy to see that it is simply *non sequitur*. From the fact that somebody wants something we cannot infer that it is reasonable to want that and what Churchland says is basically "we must hold people responsible because we need it." We all know we need it, and she is not expected to say that, but rather why it is reasonable to hold an assumption that people are responsible for their actions even though all of their actions are determined by the state of their brain when they decide to perform the action. What I want to do is the opposite – to show that there is no philosophically valid reason of punishing anybody for anything under the assumptions made by Churchland, but I will get back to it in a moment.

The second problem is how we decide what rules should be imposed on people in our societies. It is trivial, and Churchland admits it, that different people may have conflicting moral feelings about what is the right thing to do. It is even worse – a person can be in conflict within oneself that makes it impossible to decide what thing should be done or decision made. As in case of justification of punishment we can say we just need to apply rules to make it clear what we impose and what we do not. It is not philosophically very robust, but in the end it is enough that we vote or have some representatives in the parliament that will vote for us and make a decision about rules. There is no "should" here, no rules that are the right ones, just pure moral feeling of the majority of the voters that is imposed on others. One can say that sure, moral values are fully determined, and the history of each particular brain is responsible for having them, but it is not accidental that people share similar values – they share them, because they are supposed to be evolutionary adaptive (Churchland, 2012, 112). There are three responses that I want to say to that. First is that even if these mechanisms are evolutionary adaptive, it is still enforcing something on other people only because they were differently determined during their lives – it is still not a philosophical argument for doing so, rather practical. It is not a very forceful argument, but it shows a kind of shift in moral thinking – here we *use* morality and enforce it on other people because it is advantageous for *us*. And Churchland makes it explicit that we punish these people not for them, but for us – we want to have a peaceful social life (Churchland, 2006, 44). But since this is all what (at least a few) naturalists expect from morality, I will leave this argument as it is.

The second thing is that there are numerous possible non-adaptive (as opposed to counteradaptive) concepts that we can consider moral. One of them is a shocking idea I learnt about when watching a show *Star Trek: The Next Generation*. In one of the episodes the spaceship Enterprise encountered a society whose members lived only up until their 60th birthday. But these people do not suddenly die just on their 60th birthday – they are forced to commit a suicide to release resources that can be useful for younger generation. The word "forced" makes probably a wrong impression that people turning 60 in that society do something against their own will and have personal difficulties in committing suicide. Quite the contrary, the tradition made it "natural" for that people that this is what they have to do. It goes so far that in the episode in question a daughter of one of the tribesmen with tears in her eyes tries to convince her father to fulfil the tradition on his 60th birthday.

Is it something we cannot think of introducing to our societies? What is the most important here is to acknowledge that even though we can think we can try to give rational reasons not to adapt this custom in our societies (or to adapt it, it does not matter), from the naturalistic perspective the alleged reasons we give are just a result of our personal adaptive processes. If somebody says that it is immoral to enforce people to commit a suicide when they turn 60, it is just because that person's brain was shaped during their life to give such a response. Under different conditions the same person could give an argument that it is reasonable to do it, because there are more resources for younger generations which gives more opportunities for the growth of the society – such a person would be on the other hand determined to value more their society as a whole than particular lives. We may have views from different people and in the end they are worthless for the same reasons – they are just responses learnt through reinforcement and none of them is therefore correct, because there is no gold standard to adhere to. Even if one says that even without naturalism this gold standard is hard to achieve or unachievable, it is still something that as an idea can guide us when we think about the law or morality. Naturalism makes everything clear – there is no such a thing as a gold standard in morality. For every possible moral rule, the only reason why there is anybody to defend it, is because in the past this rule was profitable for their ancestors living in a society that was governed (among others) by that rule. So it makes it impossible to judge other people and societies – it is hard to treat even one's own judgement seriously when one believes it is just a matter of coincidence.

So one thing that is controversial here is the source of moral rules. Another thing is what Churchland replaced free will with. Is self-control introduced by Churchland enough to expect any specific behaviour from anybody? Even before that, what can we expect from the notion of a person that can be achieved on the naturalistic ground of eliminative materialism – or in other words, is there a place for the notion of a person that possesses self-control in this metaphysics? Churchland says:

> Is one cheapened by this neuroscientific knowledge? I think not. Self-esteem and self-worth are wholly compatible with realising that brains make us what we are. As for self-esteem, we do know that it is highly dependent on successful social interactions: on respect, love, accomplishment, but also on temperament, hormones and serotonin. (2006)

As we can see, she does not give any single argument to support her believe that "one is not cheapened by neuroscientific knowledge." Of course, she says that for example self-esteem depends on many factors that can be ascribed to a person, but she does not mention that all these things are completely independent on that person – they *have to* be determined by factors that are not dependent on that person, otherwise her whole conception of determined physical universe is flawed.

But let us make it more concrete. Let us imagine a tree that fell and after removing it we see that it imprinted Mona Lisa in the place it fell. It would be very peculiar to see it, but would we praise the dead tree for it? Probably not, we would probably say it was just an accident, even though very unlikely. Why do we praise da Vinci for his "Mona Lisa"? There may be many answers to that question. One may say that he needed to have a great talent to be able to paint this masterpiece, but I believe it would be a counterargument – if there was another person who had much less talent but was able to paint "Mona Lisa" at the same time, that person would deserve more praise, as talent (if it is a coherent concept, actually) cannot be dependent on the person that has it and if somebody with less talent can do the same thing, they probably needed to spend more time on improving their skills. But let us go further. Why should we even praise a person that spent a tremendous amount of time on improving their skills? Here we also have a quick and intuitive answer: because not everybody is willing to do that and there are temptations that make people do other things or waste time instead of spending it on things like improving skills. But does it depend on the person that decides to improve their skills to decide to do that? If, as Churchland claims, that person is just their brain and their brain is a deterministic machine, there was no way for that person to not spend all these hours on improving their skills. The compatibilist response to that is probably something along the lines: "That's true, but it was still this person who decided to do what they did, even if they were determined. They could have done otherwise, because they would have done otherwise, if they had chosen to do otherwise." This is just a standard restatement of what for a typical compatibilism free will and "could have done otherwise" mean. But let us think about one important question. Under what conditions would that person have chosen otherwise? It is clear that nothing in that person could have made a difference, unless we give up the idea of person being fully determined by brain which is a deterministic machine. There are only two options: either the beginning conditions of life of that person or the subsequent environmental influence on that person would have to be different. And those, on the other hand, are by definition something that a person cannot control. This way all of us could have done other things in our lives – but the only thing that could make it happen are outside us.

The same argument can of course be applied to things people are blamed for, so also to the examples Churchland gave. The difference between having socially unacceptable urges due to some brain tumour and without it is fairly similar to the situation when somebody can do something due to their talent or without it – in both cases only conditions outside a person could change their behaviour. The notion of self-control is therefore extremely weak, not to say right away it is meaningless. There is one difference between a person having a tumour and a so-called "healthy" person – no matter how we try to reinforce socially acceptable behaviour, a person with a tumour will not respond in a way we want, whereas a "healthy" person at least in principle can be forced to acquire "correct" responses to the urges that may cause unacceptable behaviour.

This is probably the reason why this extremely weak concept of control was invented – otherwise it is just a usual compatibilist kicking the can down the road. Here it is where the two problems I discussed meet. Churchland's aim was to provide a ground for holding people responsible for what they do and here the ground for that is that if we punish people, under some circumstances they can change their behaviour and we will all benefit from it. It does not seem to be a good philosophical foundation of a punishment. If what Churchland suggests is true, it means that in every society there are different groups of people that have different kinds of values due to having different history and genetics, and the reason why some particular values are enforced is that by accident there is a majority of people that share them. And these values are enforced on agents that on one hand could not have done otherwise, but on the other are capable of being formed in a way that in the future it will be more probable that they will act in some particular way, expected by the society.

All of this leads to I believe the strongest argument I would like to give against the view on morality that is presented by Churchland. It is an argument whose aim is to show that in the core of her idea of morality is its own denial – if morality is what Churchland believes it is, then everybody *should* act against it whenever it is advantageous to them. If my argument is correct, I believe it should be devastating for the naturalistic idea of morality, as whatever morality is, we can probably all agree that it is supposed to provide us with rules we should follow, not break on any occasion when it is going to help fulfil our own desires. It is sometimes argued (I believe incorrectly) that there are moral rules that it is morally right to break as a way to help others, probably the most famous example of which was presented in Kant's essay *On a supposed right to lie from philanthropy*. But I have never encountered any argument saying that we should break moral rules when it is advantageous for our own good – if we can say anything about morality, it is for sure that its aim is to provide reasons to sometimes act against our own good.

It is also not different in case of naturalized "morality" endorsed by Churchland. She believes that morality is a "pragmatic business," and it is about "figuring out how best to organize ourselves into social groups."[15] So let us take an example of a rule that was supposedly introduced with such an aim. One such rule can be "do not steal." So I should not steal because it will be advantageous for the whole society if we impose such a rule on everybody – I do not steal from you, you do not steal from me, and we are both happier than if we were to try to steal each other's goods on any occasion. There is nothing inherently wrong with stealing, evolutionary oriented philosopher would say that it is a natural instinct for an organism to want to have more because it can help with reproduction, which is always the ultimate goal from evolutionary perspective. If I try

---

15 One of the problems with this approach is that it is not clear where we get the meaning of what "best" is, but this I have already discussed, and here I am going just to assume we have any rule that conforms to what is best for the society.

to steal from you, however, I risk being excluded from the society. But let us think about a particular situation – what if I can steal something from you and I am sure it will never get revealed? Can I do that? According to the rule I cannot, but according to the reasons why we impose these rules I should. I should, because the only reason for me to follow these rules is because I can get punished. It does not help to say that in the first social groups humans introduced rules of conduct to increase each other's probability of reproduction. In contemporary societies it is highly improbable that helping anybody will pay off in that way (or probably any other) and therefore it should not be any reason to act upon. Also any kind of charity is an absurd atavism from evolutionary perspective. It is hardly imaginable that it could help either the society or the donor.

It also does not help to say that because of the moral instinct built in my brain I actually cannot steal anything from anybody. If it were the case, I would not need any rules imposed on me, because with or without rules it would be impossible for me to steal something. One could say that it is not really the built in moral feeling that I should not steal that causes me not to steal, but it is the built in respect for rules that in combination with a rule "do not steal" causes that, but it is also clearly false, as not every rule works like this – if somebody introduced a rule "do not throw away food even if it is spoilt," there would probably be only a few people that could conform to it and of course there is an infinite number of arbitrary rules that we can introduce and hardly anybody will follow them. One could say that it is the combination of my moral instinct that stealing is bad and the moral rule "do not steal" that cause me not to steal, but it is even more far-fetched. The point is that the rules are not for those who can only obediently conform to them, but for those that can at least on some occasions break them.

It is peculiar that one argument that Churchland gives against utilitarianism strongly supports this conclusion and acts against her own notion of morality. She makes a point that since "ought" implies "can" and since mother and father cannot leave their two children to take care of twenty other children (because their biology makes it impossible for them), we cannot say they ought to do it. But this is what a utilitarian should argue for, since it would increase the amount of happiness in the world. Let us leave utilitarianism aside and think about how the conviction that "ought" implies "can" works with the naturalistic view on morality endorsed by Churchland. A person that does something against moral rules present in the society they live in clearly cannot abide to these rules for the same reason a mother cannot leave her children – their brains make them do so. If a person murders somebody, they could not have not done that and therefore we cannot say they should have not done what they did, in other words, they were not obliged to not kill that person.

All of this leads to a conclusion hard to accept that from evolutionary perspective it is the most profitable and rational to function as if one conformed to the rules, whereas in reality break them on any possible occasion[16]. With this perspective in mind I think we can divide people into three groups. The group that is in the worst position are people that do not abide to rules but cannot hide it and are easily caught. These are the people that steal, rape, murder, but are not physically and (mainly) psychologically strong enough to escape our tribal justice and go to prison. The second group is probably the most numerous group in the society – these are all people that either believe moral rules should be abide to (because their brains make them think so) or people that wish they did not have to abide to rules, but they are afraid of being caught and punished. And the last group are all the people that do not abide to rules and are skilful enough to not get easily caught. These people are in clearly the best position from evolutionary perspective. To give a concrete example, a man that rapes multiple women, especially in countries where abortion is illegal in any case, has a great chance of his genes being reproduced. He does what he does because his brain causes it, he has no compassion towards his "victims" (how people from the second group call them), because this is just how his brain works. People from the second group call such people "psychopaths" or "sociopaths," but from evolutionary perspective they are just the most advantageous group in any society. The

---

16 There is a problem with the notion of being rational in this conception, because in a world in which eliminative materialism is true, what is called "rational" is also determined by reinforcement and this way can be changed. So when I talk about being rational here, I do it (and I have to do it) from the perspective of a person outside this system, otherwise what I think is rational would also be caused by reinforcement, and the whole argument would be circular.

question is whether the second group can reasonably impose the rules they abide to the other two groups. I believe what I have said so far makes it clear that the answer is no. The only reason why the second group imposes some laws on other groups is that it is convenient to that group – people from this group have to abide to these laws and imposing them on others can only make people from the second group be better situated. I said it is the "reason," but actually it is the cause why that second group behaves this way, as all reasons are contrived natural causes in the view of eliminative materialism. But to say that there are people that do injustice from perspective of other people and these others punish them and all of that is because of the causal relationship in which their bodies are with their environment is barely what we want to achieve – we know that. Saying that we punish others because we are determined to do so is an excuse, not an explanation and does not count as a reason.

If that was not enough there is one more important thing that we can say about Churchland's naturalistic morality from perspective of eliminative materialism. Eliminative materialism makes all ethical deliberations completely useless, because to be able to meaningfully talk about morality we must assume it is meaningful to talk about subjects that *want* something. The very idea of morality stems from the fact that different subjects may want different things and it may be impossible to make them all have what they want[17]. But even the much less ambitious project of Churchland cannot be fulfilled without the notion of will – even such a diminished one as "self-control." To make sense of self-control we must assume there are subjects that want something. They sometimes have to suppress their temporary urges in the name of the greater good which in the end is their own reproduction. But except for making "self-control" a

---

17 It is probably not a coincidence that Kant expressed his categorical imperative using the word "want". He says "handle nur nach derjenigen Maxime, durch die du zugleich wollen kannst, daß sie ein allgemeines Gesetz werde." which can be translated as "act only according to that maxim which you can at the same time want to become a universal law." He does not say "act only according to that maxim that could become a universal law" and it is something that cannot be underestimated – moral law is where all wills of rational beings must by necessity converge.

fairly reasonable concept, to make the whole idea of naturalized morality work Churchland must assume there are others that have their will which is expressed by what they want. For example one person does not want to be robbed, therefore another must suppress their urge to steal from that person. With eliminative materialism everything gets complicated – we do not have propositional attitudes at hand any more to be able to easily express the need for moral rules. Using neural nets as a model of what happens in the mind makes it clear that the only thing we can talk about are inputs to the neural net, its weights and activations given input and the output, which can either be an input to another neural net or behaviour. If somebody steals something from me and runs away I may chase them, but the fact that I chase somebody having something in their hands does not capture the fact that it is my thing, and I did not want it to be stolen. Even more vivid example is probably lying – how can we express that somebody lied to another person without using propositional attitudes? This problem was very well illustrated by Davidson when he said that even if we luckily found true psychophysical generalizations, there would be no reason to believe it more than roughly true (Davidson, 2001, 216). It is because any physical description of an event will never have the same content as a mental description, even if they referred to the same event.

Here I will stop my analysis of free will in the light of eliminative materialism – I believe we can clearly see eliminative materialism makes it impossible to make sense of notions like "free will" or "moral responsibility." I will also refer to eliminative materialism a bit more in the next chapter to make a point what this stance can help us in to formulate a more sensible stance of free will.

Summary

The aim of this chapter was to present how materialistic stances regarding the psychophysical problem can answer the question whether human beings are free individuals. Quite the contrary to what we could see in the previous chapter, we saw that materialism gives a perspective that is easy to grasp as a whole from the theoretical perspective, but it makes much harder to account for coherent ideas of free will, morality and responsibility. Of course, there are various tricks, usually based on wordplay or redefining key terms, that materialistic philosophers try to use, but there is no convincing materialistic solution to the psychophysical problem that could serve as a basis of a distinction between entities that we can ascribe moral responsibility and moral value to and those with which we cannot do this. I believe with this respect materialism is a dead end. In the end, no matter what kind of words we use to describe a human being, on the ground of materialism on the physical level it is the same kind of entity, describable using the same laws, as for example a stone or a tree. Trying to use the notion of morality as something that can distinguish human beings from other entities is also ineffective. As we could see, the notion of morality on the basis of materialism does not make a lot of sense either – it is devoid of all important (or even "essential") features.

I believe here is the point where all things I have considered so far converge and it is possible to clearly see the problems with various stances I have discussed so far. Compatibilism combined with materialism is wrong because no matter how hard we try there is no way we can meaningfully express an idea of having control over oneself, let alone free will and moral responsibility – and I do not think there is anybody that could seriously give up on these concepts. This is what we have seen when discussing Dennett, Churchland and probably Davidson (if we consider him a compatibilist). Non-materialist stances presented so far usually have a problem with explaining the relation between the mind and the brain – either the mind and matter are treated as different substances and the relation is hard to comprehend or there is some strange idea of mind being caused by the brain involved and then it is not clear what the function of the mind could be. Incompatibilist stances presented so far usually have a problem with explaining what the difference between a random and free behaviour is. On the other hand I also said that eliminative materialism is the only coherent stance in philosophy of mind and although it is obvious that a theory can be coherent yet false, I believe there are reasons we should stick to eliminative materialism in natural sciences. In the next chapter I will argue that a version of Kantian transcendental idealism employing eliminative materialism is a minimal assumption we have to make to be able to consistently speak about science, morality and a subject that is engaged in both scientific and moral activities.

Chapter 4: Free will and transcendental idealism

Introduction

The last chapter of this thesis servers two purposes. First of them is of course to give a summary of what I have discussed so far. I had to touch quite a few different areas of philosophy in order to arrive at probably not a very interesting conclusion – that the contemporary mainstream debate on the topic of free will is highly unsatisfactory. There is of course much more to say about it, and I will deliberate on that topic a bit. The second purpose is to provide the only possible, I believe, solution to the problem of free will that makes it possible to reconcile science and morality. So before summing up, let us consider this solution.

Theoretical problems with eliminative materialism

In the last chapter I discussed the idea of eliminative materialism. I argued that it is impossible to consistently claim that eliminative materialism is true and at the same time believe in existence of any kind of moral obligations. Now I would like to get back to the critique of this stance, but now to look at it from theoretical perspective – I will analyse the arguments against eliminative materialism that do not refer to morality. The aim here is not the critique by itself. I believe that eliminative materialism is ultimately the right stance in the science of mind and this critique will make a connection with Kantian transcendental idealism possible.

At first I would like to refer to the counterarguments that Paul Churchland gives to arguments against eliminative materialism in his article *Evaluating Our Self Conception*. Churchland discusses five arguments and two of them are especially worth discussing. First of them is an argument that eliminative materialism is self-defeating. It is probably very easy to conceive quickly after one gets familiar with eliminative materialism – for me, when I encountered eliminative materialism, it was also very peculiar how somebody may claim that someone claims that propositional attitudes do not exist, but yet to write a book about it. In the core of this argument is that since we form our thoughts using a language with propositional attitudes, it is impossible to argue for eliminative materialism, because any formulation of this stance involves propositional attitudes. This is what Churchland calls a conflict between the eliminativist's apparent belief that folk psychology is false and his concurrent claim that there are no beliefs (Churchland, 1993, 214).

Churchland gives a very interesting response to this argument. According to him, logically the situation is obvious – if we assign the framework of folk psychology assumptions to Q, then the proof goes like this[18]:

1. Q
2. Q → ~Q <Q and other empirical premises to the conclusion that not-Q>
3. ~Q ∨ ~Q <Material implication>
4. ~Q

So for Churchland it is just a simple *reduction ad absurdum* where starting with the assumption that Q we arrive at the conclusion that not Q. He believes that if the "self-defeating" objection were correct in this case, it would be a signal that all formal *reductios* are incorrect, since they all presuppose what they try to deny and this would be a major contribution to logic (Churchland, 1993, 214). Now the question is whether the way in which Churchland arrives at eliminative materialism is really a *reductio ad absurdum* from the assumption that folk psychology is true. Let us compare his reasoning to a clear *reductio ad absurdum* and there is no better place to find a clear case of this type of reasoning than mathematics. A simple *reductio ad absurdum* can be used to prove that there is no smallest positive real number.

1. Q – r is the smallest real number <assumption>
2. r/2 is a real number greater than 0 and smaller than r <algebra>
3. ~Q

Since one may say that it is different because mathematics is not an empirical science, let me also evoke a very simple (and formally exactly the same

---

18 Here and in the subsequent pages I use the notation used by Churchland in his article *Evaluating Our Self Conception* (1993, 214).

as the argument against eliminative materialism) *reductio ad absurdum* against existence of phlogiston.

1. Q – phlogiston exists.
2. Q → ~Q <Q and other empirical premises to the conclusion that not-Q>
3. ~Q ∨ ~Q <Material implication>
4. ~Q

Is there any difference between these three reasonings? I think it is very clear that the last reasoning is forcefully fit into the pattern of *reductio ad absurdum*. We do not really need to assume that phlogiston exists to prove that it does not exist – it is the empirical premises that do all the hard job in the proof and without the first premise everything works in the same way except for the fact that we do not arrive at any contradiction. But arriving at a contradiction is not a value by itself. We can formulate an infinite number of proofs to prove obvious existential claims using *reducito ad absurdum* in the way Churchland uses it. For example we can prove that elephants exist.

1. Q – elephants do not exist.
2. Q → ~Q <Q and empirical proof that there is at least one elephant>
3. ~Q ∨ ~Q <Material implication >
4. ~Q

Stating the first premise does not give any interesting information and does not make it possible to prove that elephants do not exist. What about the proof that there is no smallest positive real number? Here the situation is quite different. To make this proof work we need to assume that there is the smallest positive real number so that in the next step we can use our knowledge of real numbers to prove that if that particular smallest number exists and we pick it, we can pick a smaller positive real number. I believe we can easily see a big difference between this proof and the alleged *reductio ad absurdum* used by Churchland. He did not really have to assume that folk psychology is true to proof that it is not true. The contradiction is in a completely different place – he needs propositional attitudes to state what theory in which he believes and he cannot give up on using propositional attitudes even after making his statement. Probably instead of saying "I believe eliminative materialism is true." he could say what is happening in his brain when he is saying that sentence, but it would never be equivalent of saying that he believes eliminative materialism is true. I believe Churchland used a form of *reductio ad absurdum* to state his argument only to have something firm to hold on to when discussing why it's nothing strange that he uses something existence of he tries to deny. But this is highly far-fetched. Another thing is that it is not clear if Churchland can use *reductio ad absurdum*, since it is a part of the very framework he wants to replace (Lockie, 2003, 579–580).

When discussing the same argument Churchland also tries to show that if there were a predecessor of folk psychology, then folk psychology would have the same troubles as eliminative materialism to state its point (Churchland, 1993, 214). He does not give an example of what the theory could be, which is by itself quite peculiar, just says that that theory uses "gruntal attitudes" instead of propositional attitudes. What are "gruntal attitudes" is also not explained, so we end up with a not very meaningful thought that there could be some theory before folk psychology that folk psychology could have problems with replacing because the unspecified apparatus of that theory would have to be used to state the new theory. It is not clear why it would have to be the case. As Lockie (2003, 581) points out, the argumentative and expressive power of the theory being replaced by folk psychology must contain features that are necessary to state that theory – it must be possible to represent what a theory is, what is rational acceptance, what are transcendental arguments, etc. – and it is hard to see what it would not represent that is already represented by folk psychology. Also, since Churchland imagines a group of people using "gruntal attitudes" that contemplate about their current conception it looks like the very notion of propositional attitude is present in their theory, albeit maybe without a direct reference to that concept. But later when discussing this issue Lockie makes a more general and important point that the problem with eliminative materialism here is that its claims mean that it will have to be vindicated after the new theory is fully constructed, but also what it means to do that will also have to be explained. Baker (2004, 401) uses comparison to the Neurath's ship metaphor to show the problem here – in case of eliminative materialism the whole ship has to be rebuilt in one go when sailing, but also without any replacement material.

The second counterargument discussed by Churchland that is worth mentioning here is a bit similar to the one that has just been discussed. According to Hannan, a reason to be against eliminative materialism is that there is no existing alternative to folk psychology and until such an alternative shows up, we should stick to folk psychology. Churchland agrees with the fact that without a possible replacement there would be no reason to leave folk psychology theory. But, not surprisingly, he also claims there is a very good replacement for folk psychology – connectionism. I do not think he correctly addresses the Hannan's doubts. What she seems to expect is not to say which theory will replace folk psychology, but rather to have an alternative to folk psychology that is its complete replacement. For example, instead of saying that there will be a language that will not contain propositional attitudes, Churchland should present this language and show how it is supposed to work. As we could see in the previous passages, without filling in the gaps the talk about eliminative materialism is not very meaningful. Hannan analysis the claim that the conceptual scheme using propositional attitudes makes it possible for us to claim we are rational beings is irreplaceable and makes her own, milder one: according to her, it is possible to replace this conceptual scheme, but we should bet that it will not happen (Hannan, 1993, 172). In general, this argument is just a milder version of the previous argument, and it is advisable to think why Hannan decided to form it.

The reason why Hannan claims propositional attitudes can be dispensable is that they are part of a specific conceptual scheme. After Quine's *Two Dogmas of Empiricism* it became harder to believe that there are some notions, like propositional attitudes in this case, which have to be a part of any possible theory of the world. We use propositional attitudes, and they are a central part of our conceptual scheme, but Churchland wants to believe that there are other conceptual schemes that do not use propositional attitudes and Hannan does not see a reason to say *a priori* that it is impossible. Allowing a possibility of a conceptual scheme without propositional attitudes may seem to be a small difference with respect to claiming it is not possible to have such a conceptual scheme, but in fact the difference is huge, and I believe this is what the whole discussion is about. The mere possibility that we can get rid of propositional attitudes means they are not necessary parts of describing human activity (even our own).

The problem becomes even more severe if we believe in what Davidson said: "[...] when someone sets out to describe "our conceptual scheme," his homey task assumes, if we take him literally, that there might be rival systems." (1973a, 5). So, if Davidson is right, even if we are not able to provide a conceptual scheme without propositional attitudes, the possibility of such a scheme has to be assumed by mere claiming that what we use is a scheme with propositional attitudes.

According to Davidson (1973a), we are not even able to make sense of an idea of a conceptual scheme. The reason for that is that it is impossible to find an intelligible basis on which it is possible to say that the schemes are different (1973a, 20). The bases that Davidson considers are a complete failure of translation of one scheme to another, a partial failure of translation of one scheme to another and a common ground for comparison of two schemes. But the most interesting thing four our purpose Davidson says about the interdependence between a belief and meaning. As the basis of understanding of another person we must have at least a notion of accepting a sentence as true (Davidson, 1973a, 18). When somebody utters a sentence, our assumption is that he/she holds the sentence true[19]. From attributing of holding a sentence true we have to somehow get to the meaning of a sentence and what belief of holding it true represents. How this happens is described by Davidson in more detail in his articles *Radical Interpretation* (1973b) and *Belief and the Basis of Meaning* (1974). But at the end of the latter, he gives an interesting account of why belief and meaning are irreducible to physical, neurological or behaviouristic concepts. Davidson says:

> Each interpretation and attribution of attitude is a move within a holistic theory, a theory necessarily governed by concern for consistency and general coherence with the truth, and it is this that sets these theories forever apart from those that describe mindless objects, or describe objects as mindless. (1974, 322)

So the irreducibility comes from our assumption that a speaker holds the sentence he/she utters true. Of course this assumption is often released and in many cases this is also the reason for the irreducibility – for example

---

[19] Davidson does not say that explicitly, but it seems that in this context saying that a holds p true is equivalent to attributing to a belief that p.

there are situations in which we decide to interpret some of the utterances differently than what we could infer from what they are supposed to mean (Davidson, 1973a, 18).

I believe that the consequence of all of this is quite different than Churchland would expect. First, if it is true that meaning and belief are irreducible and it is an inherent part of how we interpret utterances of others, if eliminative materialism is true, we are not able to interpret what others say – we are incapable of understanding others after the change to the Churchland's framework. But, if it is the case that we cannot make a good sense of what a conceptual scheme is, then understanding of others was always an empty (although crucial) part of the framework of propositional attitudes. When somebody says, "I believe eliminative materialism is true.", the parts "I believe" and "is true" are not translatable into the new framework and this is expected. But Churchland seems nevertheless to expect that there will be some, even if very remote, counterpart to propositional attitudes and truth predicates. We can conclude this from quotes like:

> How will such people understand and conceive of other individuals? To this question I can only answer, "In roughly the same fashion that your right hemisphere 'understands' and 'conceives of' your left hemisphere-intimately and efficiently, but not propositionally!" (Churchland, 1981, 88)

The problem that I see is that we are not even able to understand what "I believe" or "is true" mean, if Churchland is right. What he tries to do is to make an argument for a new framework using the old framework whose core elements are meaningless if the whole operation is supposed to succeed. What Churchland has to make sense of is an idea of radically different conceptual schemes that can be compared against the same empirical data. There is an important difference between saying that that there are two competing theories, and we choose one that explains data better, and saying that one of the theories we investigate is actually not even possible to be understood. The former is what Churchland struggles to say, the latter is what he probably should try to express (it involves "understanding", which of course is a part of the previous framework).

Denying existence of propositional attitudes is at its core a denial of possibility of philosophy, therefore there is no reason to pretend to treat it seriously as a general thesis anymore. Also, the existence of propositional attitudes is a prerequisite to any human activity, not only rational. It is hard to imagine how we could care about anything, including eliminative materialism, if there were no propositional attitudes (let alone that caring about something is itself a propositional attitude). Churchland seems to be overly optimistic about the possibility of transitioning to his framework and it does not seem like his proposition is really well thought out. His arguments are based mainly on analogies to situations when elimination succeeded and on the belief that in the future we will have a language that will express his ideas without actually expressing any ideas – "express" his ideas. And let us not forget about it, he believes what he says now, using the old framework, has meaning and is understandable, even though the future framework will dispose of propositional attitudes, which make things like meaning and understanding sensible.

There are suggestions that because of the various difficulties with disposing of propositional attitudes, we may want to retain both propositional attitudes and eliminative materialism. It sounds contradictory, but the idea is very simple, and it is best explained by Baker, one of the proponents of this possibility:

> Instead of supposing that the resistance of the common-sense conception to accommodation with scientific theory robs the common-sense conception of legitimacy, we may take the common-sense conception to be practically indispensable, even if, strictly speaking, it is false. (Baker, 2004, 413)

I do not find this idea philosophically appealing, to put it mildly. It reminds me of Dennett saying to the scientists not to tell people they do not have free will, because it may change their behaviour. This "philosophical denialism" is very dangerous for the very practice of philosophy. After all, if we know the philosophical consequences that we want to achieve and nothing can stop us from getting to them (we can always say we need to treat some concepts as valid, even though we *know* they are not), then what is the reason for practising philosophy? Philosophers that are not able to manage practical consequences of their theses should probably not utter them in the first place.

Another way, which I am going to take, to defend substantial meaning of propositional attitudes for subject's activity is what I call Kantian-Wittgensteinian approach to the idea of a subject. According to a well-known quote from Wittgenstein's *Tractatus Logico-Philosophicus*, the subject does not belong to the world, but is a limit of the world. Kantian perspective is best presented in Transcendental Doctrine of Method, a part of *Critique of Pure Reason*. An interesting thing is that even Churchland acknowledges this perspective as an alternative to his eliminative materialism. In his *Evaluating Our Self Conception* we can read "It is still possible, perhaps, to argue for some kind of Kantian inevitability about the framework features of FP" (Churchland, 1993, 212). What is also interesting, is the way he describes people that adhere to this idea:

> If one is thus, shall we say, a Child of an Earlier Era, this may seem Palaeolithic and regrettable to some of us, but it is not bad faith for such a philosopher to insist on some special epistemological status for FP. (Churchland, 1993, 212)

Reading this literally would give us no choice but believe that for Churchland (at least some of) the important philosophical beliefs are a matter of fashion or taste and we have to accept that some people are different in that manner than others. Although it is not of any importance to me whether somebody calls me "a Child of an Earlier Era" or calls my philosophical beliefs "Palaeolithic" or "regrettable," as these are of no substantial importance in philosophical investigations, but the previous claim is dangerous for importance of philosophy itself. Therefore, it is necessary not only to justify this Kantian-Wittgensteinian approach, but also to discuss the reasons why philosophers like Churchland do not adhere to it, even though they believe it is still an option. Let me start with a more thorough description of the stance I want to defend and then present what I believe are the reasons for its rejection by Churchland and other naturalistic philosophers.

Transcendental idealism and free will

One of the most famous Kantian ideas was to divide knowledge into knowledge of phenomena and knowledge of noumena. The latter is something we cannot have in this life, but the idea of it is crucial for Kantian transcendental idealism. I will now discuss this stance, but I will not provide a thorough description of it – too much has already been written about it and my focus is of course on how it is related to free will. Also, I will discuss what features of Kantian approach to morality are impossible to retain under the assumption of eliminative materialism. I do not see it necessary to discuss at length Kantian ideas and make separate argument for each of them, because I believe enough has been written on that matter so far – it is much more valuable to show what we lose if we accept naturalistic view on humanity, free will and morality. Therefore, in the next paragraphs I will bounce back and forth between presenting how I see Kantian ideas and what it means to accept naturalistic conclusions instead of them.

The most important thing is probably that the very reason of introducing transcendental idealism for Kant was to provide a foundation for a belief that human beings are free. It is not popular to think about Kantian philosophy in this way in contemporary (mostly naturalistic) philosophy. *Critique of pure reason* tends to be presented as mere epistemology, as if *Transcendental Doctrine of Elements* was its main part. Why the distinction into *phenomena* and *noumena* was introduced is best presented by Kant in one of the footnotes in the preface to the second edition of the *Critique* (Kant, 2007, 23). He says that the distinction is assumed so that it is possible to check whether it helps reason to avoid being in a conflict with itself. If it is so, which Kant tries to prove in the *Critique*, then the distinction is valid and has to be retained. Conflicts of the reason with itself are called by Kant "antinomies of pure reason," there are four of them and the third antinomy concerns the free will problem. The thesis of the third antinomy is that except for the causality of nature there is also another kind of causality that we need to assume to explain what happens in the world of *phenomena* and that causality is causality of freedom. The antithesis to this is that there is only one kind of causality – causality of nature. The distinction into *phenomena* and *noumena* helps to solve this antinomy by ascribing causality of nature only to *phenomena* and therefore making room for causality of freedom in the world of *noumena*.

So far so good, but if Kant was right, why would anybody reject this distinction and return to descriptions of events only from one perspective? I guess there may be many reasons for that and probably it is a question that should be directed towards philosophers like Churchland, who on one hand claim Kanatianism is a reasonable (tough outdated) stance in the conflict, but on the other suggest their own, completely different stance. But it is still valuable to think about the merits of Churchland's eliminative materialism in comparison to Kantian transcendental idealism.

One of the main problems of Kantian stance is to explain how it is possible that in the world of appearances we can spot activity that does not come from this world. Kant says we need to assume this causality in order to be able to explain all the appearances, but everything that happens in the world of *phenomena* can easily be explained by other *phenomena*. I will get back to this later, but at first glance this is one of the reasons why stances like eliminative materialism are proposed. Churchland does not have to refer to source of causality other than the causality of nature. For Kant it is not important to say how it is possible that causality from freedom has its importance in explanation of the behaviour of appearances – it is enough for him to show that there is such a possibility, which means that there is no conflict between nature and causality through freedom (Kant, 2007, 479). It is also important to note that the source of causality in Kantian conception is moral law. We are free and rational beings insofar as we act upon laws that we present to ourselves through maxims that are universally obligatory. The criterion for such maxims is compatibility with the will of every rational being. An important idea is that we have to be able to act upon such maxims no matter what our personal goals and motivations are.

Here we already arrive at a point where there is an easy and common naturalistic way to diminish importance of moral law in human actions. Kantian idea of moral law demands that we are able to fulfil our moral duties even if we have no inclination to do it. I help a beggar by giving him/her some money or food, even though I do not feel like doing it at all. The easy response is that I still have an inclination to do that, it is just unconscious. It can either be instinct built in in my biology or I am compelled to help a beggar by some social programming or something else is the foundation of my inclination, but in the end it is not that respect for moral law is what is the basis for my action. I have already discussed this kind of arguments in the previous chapter. This kind of thinking is disastrous for morality and forcing this kind of visions should end up in a complete disrespect to any rules we can be forced to obey but do not comply with our own needs. Our acknowledgement of unconscious motives in "moral" actions should make us more cautious and think about what is best for ourselves instead of playing moral games imposed on us by biology or society. The fact that we, as a society, do not want to give up on morality, criminal law, etc., means we believe in possibility of actions based solely on the feeling of respect to moral law.

A part of that belief is recognizing human beings as ends in themselves. We respect the will of others and freely make a rational decision to act in accordance with laws that agree with the will of every rational being. Even when we decide to act upon our desires, it is what we rationally choose, rather than being impelled by them.

Of course, a dogmatic naturalist may say that it is not how the things are and whenever we choose to do something, there is nothing but our experience and biological constitution that can influence our choice – and the kind of freedom Kant endorses is an illusion. What would change to the belief that human beings are ends in themselves in that situation? I believe it is quite simple – in naturalistic metaphysics there is no way to present human beings as ends in themselves and eliminative materialism is the most vivid example of how different its conception of human choices in comparison to discussed Kantian ideas is. To make it concrete let us think of a neural network that would be an artificial representation of a human being making a personal decision or deliberating about the right choice about something that will affect the whole society – for example whether abortion should be legal or illegal. Every decision is made using language, so the right representation of the situation in question will be a neural network that represents a language model. A language model is a model that captures distribution of sequences of words in a given language. There are many possible applications of a language model and here I will focus on just one – sequence generation. Here, a language model gives a possibility to assign a probability to any sequence of words. The simplest way such a model can do it is by multiplying probabilities for each word in a sequence, where the probability of each word is computed as a conditional probability of the word given all previous words in the sequence. In language generation a language model can generate subsequent words on the fly, given some particular beginning context, question, etc. So an, admittedly simplified a lot, model of processing that happens in a human brain when somebody deliberates about some problem to solve can be specified as an answer to a question that is generated using a language model trained on some particular text data. So, for example, we can imagine an input question: "Should capital punishment be allowed?" and an answer generated by a neural net could be "Yes, there are some crimes for which death is the only just punishment." or "No, right to life is a human right and no one can be disposed of it under any circumstances." In both cases the output is deterministic, and the drastic difference can be only caused by three factors: initialization of the weights in a neural network, training parameters and the corpus of texts used during training. How does this translate to what happens in a human brain? Having in mind that what I have just described is just a model, we can say that the initialization of the weights happens during ontogeny, when human brain is formed, training parameters can represent the environmental conditions that influence what a person pays attention too and the corpus of texts are all sentences perceived by that person during his/her lifetime before answering the question[20]. Let us now analyse consequences of this view.

One thing to spot immediately is that there is no part on which a person can have influence when making their judgement. It is hard to inject here even a notion of free will that Dennett defends so eagerly, namely the notion of being free in choosing what a person wants. By recursively tracking all choices of a person we get to the first one which is influenced by exactly the same conditions as all other choices. The very first choice made by a person influences the subsequent ones in a way that the feedback that a person gets from the environment can be different. But we can barely call it dependent on that person as it is a product of factors that do not depend on a person making a choice at all. The difference between two people having such a different opinion about the capital punishment is grounded solely in the difference in factors that they do not control. Moreover, and more importantly now, their opinions are just sequences of words that were assigned the highest probability after they heard the question they were supposed to answer. In case of the first-person, after that person heard the question "Should capital punishment be allowed?" their neural net assigned the highest probability to the word "Yes", then, conditionally on the sequence "Yes,", the highest probability to the word "there", then,

---

20 It is not easy to decouple the corpus of texts the human's brain is "trained" on and the influence of others on the training as it also happens vastly using language. But it should not make a big difference in the argument.

conditionally on the sequence "Yes, there", the highest probability to the word "are" and so on until reaching the sequence "Yes, there are some crimes for which death is the only just punishment." It is hard to see how such a process can contribute to our thinking that a person is an end in itself and we have to respect judgements of others when they are produced in this way.

What I have just presented is an extremely simplified model of human speech generation, but I do not see any reason to think that when more sophisticated models come into play there is a qualitative difference. In the end, no matter what happens in a neural net, the core of its working is the same. This is why we need a different source of causation to account for the difference between human beings as ends in themselves and human beings that we perceive using our senses. In Kantian philosophy it is moral law that serves as this source of causation. Moral law needs to meet a few conditions to be useful in the investigation of human free will. Moral law cannot come from the outside to bound a human being to do what it tells, because in that case will would still not be free. Moral law and free will must be connected analytically, and this is how things are according to Kant – to have free will and to operate in accordance with categorical imperative is the same. Categorical imperative is a formal way to point us at maxims that agree with free will of any rational being and therefore at the moral law.

An important thing is that moral maxims have to be *a priori* judgements, because *a posteriori* judgements can only tell us how things are, not how they should be. The neural net example from the previous paragraphs is also a good illustration of naturalistic mistakes in that regard. People who, based on their experience, say how things should be actually say in a very convoluted way how things are. Saying "Right to life is a human right and no one can be disposed of it under any circumstances." is just a statement of how things are according to the person that says it, based on their experience. It is therefore not surprising that a person with different experience can have a totally different view on the same matter and we have to remember that having a view on some matter means just producing a particular sentence in response to some question[21].

---

21 It could be more complicated, because we usually say people have their opinions

Here we arrive at another very important problem regarding the relation between free will and morality. Let us assume that Kant is wrong and human beings are not free and there is not such a thing as a universal moral law. This is what we get pretty much for free on the ground of eliminative materialism. What is the basis of imposing laws on people in our societies? One can say there are laws that are conventionally agreed upon, because most of the people believe they are right to be imposed on everybody. But, as we can see, this does not make any sense for two reasons. First, what people believe is a matter of factors that are totally out of control of these people. The conditions in which people live are not just important factors that influence their views, they are all these factors. Therefore, the laws imposed on a society are always only a reflection of this society's experience combined with people's genetic conditioning – but what matters is just the views of majority of the society. The minority has to obey laws they do not approve of, just because their experience in life and genetic conditioning were different. It is a highly paternalistic view on morality, which is nothing wrong on its own, as morality has to be paternalistic, but with combination with contingency of morality it is just an unjustifiable coercion.

Another problem with viewing morality this way is making the notion of moral progress a complete nonsense. It could be said that it is good that in many parts of the world slavery is no longer legal and it is an instance of a progress in our moral views. Is it so? To make sense of a concept of moral progress there must be a moral standard that humanity must adhere to, but it was not obeyed for at least some part or in some places of human existence. If connectionists are right, then the way people view morality in some particular time and place in the world is the result of what the neural nets they carry in their heads converged to. It is not surprising that when time and place are different (and hence the conditions in which neural nets are trained), then the result of the training is different, and it is something that cannot be judged as right or wrong. If I train a neural

---

even if they do not say them. But since having an opinion is a propositional attitude and eliminative materialism disposes of them, we do not have to be bothered by this.

net to recognize cats and dogs in the pictures, I need to have something that is called the "ground truth" about what it means to be a cat or a dog in a picture – a collection of pictures labelled as "cat" and "dog". It is hard to imagine what kind of ground truth could be learned by a neural net that tries to distinguish between "right" and "wrong" and removing the notion of propositional attitudes makes this task virtually impossible. Let us think how the training of such a neural net could look. To make it simple, let us consider only recognizing in the pictures people performing actions considered wrong and the class of wrong actions consists of thefts and murders, whereas the class of non-wrong actions is some more or less random collection of pictures of people doing different neutral things. A naive approach would be to split this dataset randomly into two datasets – training and testing – train the neural net on the training and then evaluate its performance on the test dataset. But this approach would not let us check whether the neural net learned what the concept of doing something wrong is. It could learn to associate pictures of a theft and pictures of a murder with the word "wrong" and other pictures with the word "neutral" without capturing what it is to do something wrong. A better approach would be to first group the pictures from the "wrong" group into two subgroups – thefts and murders. Then, one of these groups should be used in the training set among "neutral" pictures and the other should be in the test set, also among "neutral" pictures. Even without doing this exercise it is hard to imagine that a neural net could learn anything sensible using this approach. If, by any chance, it manages to do better than chance on the test dataset, it would probably be because of some errors in the process, like using pictures taken at night in the "wrong" part of the dataset and pictures taken during the day in the "neutral" part – in such case the neural net could learn that dark scenery is associated with doing something wrong.

It is completely different than in case of recognition of cats and dogs. In this case, a neural net learns visual patterns to distinguish between these two classes – for example, the length of the tail, shape of the ears, etc. But, as someone may rightfully complain, it is not clear that just by the look of an object we can recognize whether something is a dog or a cat. Robotic dogs (rogs) or wooden dogs should not be considered dogs. A naive answer to that complain may be that the neural net recognizes a shape of a dog instead of a dog, but it does not capture the whole complexity of the problem. In our world dogs (I presume) are only biological organisms and if we assume possibility of machines like rogs we must say that in most cases when we see a moving object with a shape of a dog we must assume it is a dog – we are not able to go through the whole process of distinguishing dogs from rogs. This is a reminiscence of the epistemological problem of knowledge, but it is very relevant in our discussion. We must know what a neural net learns in order to be able to say whether it learned it correctly. It is not enough that a neural net distinguishes objects in two pictures, but it is also necessary to know why it does what it does. It is hard enough to make an argument that a neural net can know what it is a dog and a cat, but it is surely impossible to justify why a neural net can distinguish wrong and not wrong actions.

Why? Because one of the outcomes of using neural nets as a model for how human beings process information is that there are no propositional attitudes, and moral notions are inevitably bound to propositional attitudes. It is the fact that two people may want two things that are contradictory with each other that is the reason why we use the notion of morality. If all the people were always doing things that are compatible with what others want, no morality would be needed, and no law would have to be imposed – it is not a coincidence why Kant tried to grasp the notion of morality through the notion of will. But in the world of eliminative materialism there are no propositional attitudes, there are only actions and reactions. Stealing an object may cause a reaction of its owner, but if we can track everything that happens during this event without a reference to propositional attitudes, it begs the question why we should not allow it. Normally the reason why we do not allow it is that the person that is the owner of the object does not want it to be stolen. But, obviously, we cannot make a reference to what the owner wants here. This problem of course also holds in case of the earlier discussed problem of slavery. Slavery should not be allowed not because it is customary in our culture to forbid it, but because it violates the will of a slave.

It is easy to see that in case of any action that is considered a crime, its qualification would change if a supposed victim agreed to what happened to him/her. We cannot make a rule "Stealing things is allowed as long as the owner of a thing to be stolen agrees for that," because it would not be stealing any more. This is something completely overlooked by the proponents of eliminative materialism. Paul Churchland naively thinks that connectionism can help us better understand human beings and improve social interactions (1993, 219), but he seems to forget that to improve something, the thing to improve must first exist.

These are all important reasons to not treat eliminative materialism as a stance that is enough in isolation from any context that makes it more sensible for example in context of morality. It is not an option to accept eliminative materialism and say that even though propositional attitudes do not exist, we are going to live like they did, as it would be adhering to the worst kind of philosophical denialism and if someone wants to do it they do not need philosophy. I believe eliminative materialism is a great example of why we need the Kantian distinction between things in themselves and appearances. I will now turn to a discussion of this idea.

Things in themselves, appearances and eliminative materialism

I believe the answer to the question posed in the title of this subchapter is simple – eliminative materialism is not wrong; it is just a stance that refers to human beings only as we see them in appearances. It is admittedly more complicated to justify a shift towards this stance. It is not only about the reasons why this distinction is necessary, but also about trying to understand why philosophers like Churchland claim it is a relic of the past and they moved to another paradigm. I will tackle these problems one by one, but I believe answering one of them will shed considerable light on the other.

The problem of free will is basically a problem of how human autonomy can fit in with the laws of nature that are supposed to describe our world. What compatibilists tried to do (poorly, as I have already presented) was to change the notion of free will so that it fits their rigid definitions of determinism and laws of nature. But there is another possibility, a possibility that was overlooked by them, namely, to look more suspiciously at the way we think about the laws of nature and admit that this rigid view we have of them is a product of our minds. I have already discussed various views on the laws of nature in the second chapter, when discussing possibility of violations of laws of nature. We could see there that for example Swinburne claims that we can retain a law of nature even if we perceive exceptions. Let us now discuss how Kantian ideas change the way how we perceive the laws of nature.

At the core of how the distinction into things in themselves and appearances works are categories, and from the perspective of our discussion the category of cause is the most important. It is well known that Kant builds on Hume's critique of the notion of the cause, but it is not clear that the way Kant improves on Hume's results is appreciated enough. In *Prolegomena* Kant makes an explicit reference to Hume when he is talking about the relation between categories (the pure concepts of the understanding) and experience. Hume's mistake, according to Kant, was that he never thought of reversing the way categories and experience are connected – Hume tried to derive categories from experience, but Kant claims that it is experience that is derived from categories (Kant, 2004, 64). The notion of cause could not be derived by experience, because we never experience a necessary connection between two things called by us "cause" and "effect." The connection between the cause and the effect is necessary, but any particular law is derived from experience (is synthetic *a posteriori*), and therefore has to be accidental, which agrees with Humean critique of the notion of the cause. But even Hume believed the effect must necessarily follow its cause and if we were to stick to his explanation we would not have a good reason to believe in that necessity and it could be a reason to abandon the notion of cause altogether. We might say that there are no such things as causes and the only thing we perceive is that so far one class of events has always followed another class of events. When there are any deviations perceived we can start analysing correlations, but in the end we can never come up with a sensible notion of a cause. Kant's solution lets us retain the notion of a cause for the "price" of its applicability only to appearances.

The fact that categories apply only to appearances lets us explain how it is possible that human beings can be treated as both causally determined by the laws of nature and free of that influence. But we can transfer this finding to the relation between eliminative materialism and propositional attitudes. As appearances, human beings do not possess any propositional attitudes and as scientists we have to explain their behaviour without any reference to propositional attitudes. But when we treat humans as beings involved in moral relations, we need to acknowledge propositional attitudes as an essential part of it. Disposing of propositional attitudes altogether does not sound plausible not only because it would be impossible to state what eliminative materialism is without propositional attitudes – after all, Churchland believes it may be only a temporary flaw – but because all our activities would become nonsensical without propositional attitudes. Morality, on the other hand, would become utterly impossible. What I want to say here, is that Kant gives us a much more reasonable foundation for using propositional attitudes than what we could see before – for example, Baker suggested that we use propositional attitudes anyway, even though an assumption of their existence is false. There is clearly no reason to do that and even stating such a demand presupposes existence of propositional attitudes – someone must think for some reasons that propositional attitudes are practically indispensable. So propositional attitudes seem to be valid only when we consider human beings as moral subjects and things in themselves. Although practical indispensability is also why it is reasonable to keep on using propositional attitudes, this approach is in an important way different than what Baker suggested. What needs to be acknowledged is that experience gives us only one way of seeing a human being and it is in no way more privileged than when we see a human as a moral subject. This seems to be not an option to naturalists of Churchland's kind and it's the reason why they are trying to derive the moral discourse from experience.

This conclusion should make clear what is the answer to another problem emerging from Kantian philosophy that is relevant to the distinction on appearances and things in themselves – are there two different worlds with different subjects or there is only one world and the distinction in question lets us think about two different aspects of that world? It is astonishing to me that there is a discussion on this matter, although I believe it is mostly a dispute based on the fact that Kantian stance is not easily explainable using language that after all emerged during our usage of the theoretical reason. For example, what do we mean by the word "world"? Does it refer collectively to the appearances or to the "things in themselves"? It is quite obvious that it will have different meanings when we talk about two different worlds with different subjects and one world with subjects with different aspects depending on whether we think of them as appearances or things in themselves. What is important is that it is unreasonable to question the identity between the empirical and noumenal subjects, because this identity is the main reason why the distinction between phenomena and noumena was introduced in the first place. Questioning identity between some particular person as he/she appears in experience and as a thing in itself is like questioning the identity a = a.

But it does not make it obvious that the two aspects account lets us solve the free will problem so easily. For example, many consider Davidson's anomalous monism a two aspects account of Kantian stance (e.g. Nelkin, 2000, 565). I have already discussed why it does not reconcile determinism and free will in a way Davidson expected it, let alone meeting Kant's expectations, so just to briefly give a context it is enough to say that when we identify mental events with physical events and say that an event is determined under a physical description, it does not matter at all that it may not be determined under a mental description. In general we can say that if an event has a description (does not matter if it is mental, physical or whatever kind of description we may think about) under which it is determined, then it is determined as an event, no matter which description we use.

To make it clearer, let us kind of reverse the relation between physical and mental descriptions provided by Davidson. Let us think about the world in which physical descriptions do not provide deterministic explanations, but mental descriptions do. In other words, in this world we can construct nomological laws between events based on mental descriptions, but when we describe them physically, they seem to exhibit randomness. For example, let us think about a simple law that says that if I believe there are burglars in my house, I immediately want to kill them, no matter what kind of threat they pose, and do everything needed to accomplish this. And let us assume, although I do not think it is necessary, that everybody has the same kind of reaction on the belief that there are burglars in his/her house. This lets us create nomological laws on the psychological level, but at the same time there may be no strict physical laws that bind together our physical responses. Still, if we identify mental events with physical events, we can insist that on some particular occasion the sequence of physical events is determined, and we know it through the determination of mental events.

Why is it all important? Because it highlights what the problem is – it does not lay within finding nomological laws between events, but rather within ascribing too much meaning to (or even substantializing) causal relations between events. We have to remember that as things in themselves, the events are not determined – it is our category of cause that binds them in causal relations. It is clear that Davidson makes this mistake – he identifies mental events with physical events, but these are just two different ways of describing appearances and he never relates them to the things in themselves. I believe that this kind of mistake is the root of all problems with contemporary stances regarding free will. If one believes we are just matter, matter is governed by strict physical laws and these laws are not subject dependent, then there is nothing to do any more. And for philosophers like Dennett or Churchland this seems to be the starting point.

Let me show you an example of laws in a field other than physics to illustrate my point. A while ago I heard a girl telling a story of how she talked to her Turkish fiancé about Turkish language. She was not Turkish, and she did not know Turkish, but she learnt it. She said he corrected her grammar mistake – she said something that according to the grammar rules she had learnt seemed to be correct, but he felt it was not right, although he could not give an explanation why. On her class she said about this problem and got an explanation – she was right about the grammar rules, but for some, not important for us, reason they did not apply to that case. Here comes the conclusion. She came back from the class and said to her fiancé that he had been right, but now she knows *why* he was right, because she knew the full rule. What I want to say is that the rule she got to know does not matter at all to him, a Turkish native speaker. The laws of Turkish grammar are not imposed on Turks, they were extracted from the way they use Turkish language in order to make language comprehension easier for foreigners (among some other reasons). We can generalize this to any language in general and say that the relation is quite the opposite to what appears to be the case – it is not that people talk in some particular language in some particular way because of some ground rules, but it is that these rules are formed based on how these people talk in this language.

This relation is quite a good illustration of the relation between the laws of nature and how we perceive nature in Kantian philosophy. The laws of nature are extracted by us from our perceptions of nature, which are themselves affected by our conditions of the possibility of experience – intuition and categories. To assume that the nature will always behave in accordance with the laws of nature that we perceive is useful, but not accurate. It is like expecting people to always talk in accordance with grammar rules of their language. It may be expected to some extent, but if we see them talking in a different way than what we could expect from our rules, then we may decide to change our rules or add exceptions (which actually make clear that some statement is not a rule), etc.

Here I made a short remark on a controversial topic of whether the laws of nature can change. I am not going to discuss this issue in detail, but it needs some more explanation. There are actually two different things that may change – our descriptions of laws and the laws as they are in themselves. It is quite obvious that the descriptions may change and should change whenever we experience some deviation that has to be accounted for. The question, of course, is whether the law of nature can itself change. There are philosophers that believe laws of nature are immutable, but contingent, and such an account is presented for example by Lange in his article *Could the Laws of Nature Change?* (Lange, 2008). Lange at the very beginning of his investigations in the article makes an assumption that whether a law is true or false is a feature of the world independent of a mind and it gives some clue to the boundaries of his (and philosophers like him) approach. There are two options – either his considerations apply to the world of phenomena, or he does not acknowledge the distinction into phenomena and noumena and assumes there is a world independent of the human mind that we can say something sensible about. To me it is quite clear that the second option is the right one – like other contemporary philosophers Lange seems to not acknowledge the critical philosophy of Kant and tries to make statements about the world how it is when we do not perceive it. It should also be clear that in Kantian philosophy it does not make any sense to talk about the world in this way. But this problem highlights an interesting issue that is worth investigating further – if laws of nature are indeed immutable (as Lange claims), then it is a synthetic *a priori* judgement, as we are not capable of experiencing immutability.

I will not pursue this issue here, but it is worth noting that dogmatism of Lange's kind is widespread throughout philosophy and especially in the investigations in free will, although it may be not explicit. One of the main issues that needs to be answered when somebody tries to solve the free will problem is whether the Laplace's demon can predict all subsequent states of the universe. The idea that the Laplace's demon incorporates is still very vivid in contemporary philosophy, which was very visible in the first chapter, where we defined determinism as an idea that all laws of nature together with the state of the universe at a specific point determine all subsequent states of the universe. There is, however, an important difference that is introduced when we consider determinism from the perspective of the Laplace's demon. The Laplace's demon is an abstract entity that is supposed to have *knowledge* of all physical laws and all facts regarding the state of the universe at a specific point. What is often overlooked, is that belief in at least theoretical possibility of such an entity presupposes very strong claims about the nature of the laws of physics and the facts regarding states of the universe. Namely, that there are some laws of physics that are absolute and ultimate, and our physics is either some kind of approximation or a distant ancestor of them – that our human perspective will in the end make no difference in how the laws are formulated. Of course, they may be formulated in different languages, but the descriptions must agree.

This dogmatic view is of course consistent with common sense, but it seems to not take into consideration the achievements of the critical philosophy. I call this view "dogmatic" not without a good reason – in Transcendental Doctrine of Method, in the first *Critique*, Kant calls dogmatist philosophers that try to make metaphysical statements about the world as it is in itself without acknowledging the influence of the reason on how we perceive the world. The dogmatists of his time were classical metaphysicians, e.g. Cartesians or Wolffians, that tried to fit the immaterial soul into the description of the material world. Nowadays we have a different kind of dogmatists, and these are various kinds of naturalists who have a very similar task – they want to describe the world without referring to immaterial entities. As we could already see in various places within this thesis, their problems are complimentary. Where the dogmatists of the old days had problems with explaining how it is possible that an immaterial soul influences the material world, the dogmatists of today cannot justify making a few objects of the material world responsible for their deeds, whereas others cannot be held responsible. I will not get into details here, as I have already extensively referred to both kinds of problems.

But not only dogmatists are Kant's target in Transcendental Doctrine of Method. Another group of philosophers are sceptics that try to diminish any kind of metaphysical endeavour. The most prominent of them is Hume with his critique of the notion of cause that I have already discussed. A sceptic, disappointed by constant disputes in philosophy, decides to withhold his/her judgement in metaphysical matters. The problem of Hume's scepticism with respect to metaphysics is well described by Stern in his article *Metaphysical Dogmatism, Humean Scepticism, Kantian Criticism* (2006). As he rightly identifies it, the problem of Hume is that by undermining metaphysics he eventually has to undermine his own stance and by this move he makes his scepticism an implausible stance. Which, of course, paves the way for new metaphysical disputes. What Kant tries to do is to show that the only way we can end the dogmatist metaphysical disputes is by becoming critical philosophers and acknowledging that the category of cause and other categories of reason cannot transcend the limits of experience. But there is a price for that – we are not able to make metaphysical claims using these concepts.

I believe nowadays we can observe disputes similar to the argument between dogmatists and sceptics. As I have already indicated, the dogmatists of today are naturalists. Their opponents are various kinds of postmodernists whose common trait is relativizing all the knowledge or values to the time, life conditions, personal needs, etc. It is therefore not surprising the battlefield of metaphysics is very similar to what Kant described in his times. The disputes I presented in the first chapter of this thesis show how much critical philosophy is needed in the discussion of free will, as there is no visible progress seen in the contemporary debate. Various philosophers go round in circles redefining terms and hoping it will magically help in resolving one of the most intrusive philosophical problems.

Before wrapping up, let me discuss two more things. First, I need to get back to the dispute between compatibilists and incompatibilists and show which of these stances is represented by critical philosophy. Second, it is important to refer to the accusation that Kant is a mysterianist.

Kant and free will in relation to compatibilism and incompatibilism

Not for everybody it is clear whether Kantian views on free will can be seen as compatibilistic and incompatibilistic. There are numerous and dense discussions on this subject driven by an idea of conformity to the moral laws and possibility not to do it when an agent is determined to conform or to not conform in some particular situation. As I clearly stated at the end of Chapter 1, I do not believe that the compatibilism-incompatibilism dispute can be fruitful. I believe it is even more useless to try hard to prove that Kant was a compatibilist or an incompatibilist (unless someone really wants to have Kant "on his/her side"). But the discussion is very instructive in terms of providing reasons to believe in the vision of free will that Kant endorsed and its relation to the determinism.

Let us think why Kant could be a compatibilist. Someone could argue that in the end Kant shows how determinism of nature can coexist with free will as the empirical world is governed by categories one of which is the category of cause. So, everything that happens in the empirical world is determined, has its cause in the antecedent conditions, yet we are free. But, as a proponent of a thesis that Kant is an incompatibilist can say, freedom comes from a different source of causality than the causality according with the laws of nature and therefore it shows that for Kant free will does not exist as a part of the deterministic world. This is close to a reformulation of the third antinomy of pure reason. The thesis of the antinomy is that there is a source of causality other than the laws of nature and the antithesis is that there is none. By changing the perspective of how we view human beings Kant is able to accommodate both these views and therefore solve the antinomy. So we could say that Kant is both a compatibilist and incompatibilist, but this makes no sense – the whole idea of making this distinction lays in the fact that these categories are mutually exclusive.

This shows how much different are Kantian views when comparing to what both compatibilists and incompatibilists claim about free will. I believe it is because this distinction was formed for the needs of contemporary philosophy with pre-critical mindset. In this pre-critical mindset dualism and materialism are thought of as the only available options for the solution of the mind-body problem (e.g. Searle, 2004, 117) and all formulations of the solutions to the free will problem are related to either dualism or materialism. Critical philosophy leads to a completely different space of solutions. For example, Kant's solution to the free will problem is to derive existence of free will from the existence of morality, not the other way around. Trying to fit his solution into the contemporary framework that was built to serve completely different needs is, I believe, completely futile.

Kantianism and mysterianism

Dennett argues against dualism that even if he does not have a decisive argument that would prove dualism wrong, he believes it is not an attractive stance from explanatory point of view, because a dualist must admit there is a mystery that cannot be solved (Dennett, 1991, 37). Kant is neither a dualist, nor a materialist, but there is also an element of mysterianism involved in his philosophy. One of the things we cannot know is how it actually happens that we can make a decision that is influenced by moral law, and it makes a difference in the world of appearances, even though the whole world of appearances is governed by the laws of nature. How does it compare to the solutions endorsed by a naturalist, who claim they can know what cannot be known on the ground of critical philosophy?

We have to acknowledge that naturalism cannot offer anything more than critical philosophy can. And it is because naturalism is a subset of critical philosophy – naturalists are philosophers that decline to leave the world of appearances in their work. So whatever insight a naturalist may have, it can be used with ease by a critical philosopher. If a Kantian philosopher does not endorse some naturalistic solution to a philosophical problem (like Dennett's "solution" to the free will problem), it is not because there are some tools that are inaccessible to them, but rather that the solution is not convincing. I would say it is the other way around – a naturalist must try hard to come up with an idea how to fit in naturalism some categories that were not designed for this view and free will is of course an example of them. I have never heard of "antinaturalising" any phenomenon that a naturalistic philosopher or a scientist came up with, whereas naturalization of notions is a usual naturalistic practice. It is because an antinaturalist has all naturalistic means available to him/her, whereas the other way around is not true.

What does it help us when dealing with mysterianism? I believe the answer should be clear by now. An antinaturalist, for sure one of a Kantian kind, can "know" everything that a naturalist can know. When Dennett claims he knows how exactly free will can coexist with the determinism, he changes the meaning of the term "free will" to be able to fit it in his usual, naturalistic understanding of the determinism. Let us say that this claim has a form "X can coexist with the determinism" to get rid of the dubious term. Whatever is referenced by Dennett by X can also be referenced by any other philosopher. Another philosopher may just not agree that there is any sensible reason to say that A is free will, but can agree with all descriptions of the processes that involve the notion of "free will" in Dennett's world – that there are people that do things that they want to do, etc.

So in the end, I do not think that Kantian mysterianism is in any way harmful for philosophy. It should not be surprising that one of the effects of the critique of the reason is that we can know what things we cannot know. Redefining terms will not help with anything – we will still not be able to know these things. So the "accusation" of mysterianism is as much right as it is ineffective. We can also summarize this answer in other words. Psychophysical problem is not an empirical problem and only such problems can be assumed to be possible to be solved by science.

Summary

In this chapter my goal was to summarize the idea of free will that is sensible to adhere to in the light of the considerations presented in the previous chapters. I did not see it necessary to defend extensively the Kantian idea of free will – there are numerous works on that matter. What was more important to me is why there was a naturalistic shift in the free will research and why it is futile to stick to it and these are the things that I presented in this chapter.

I believe this summarizes my efforts to show the current state of the debate over the free will problem. As metaphysics of the eighteenth century needed a critique of reason, so the contemporary thinking about free will needed a detailed critique of what we can achieve and whether the current path can give any tangible results. A metaphorical way to describe the state of the contemporary free will debate is to say it is "walking in circles." It was best visible in the first chapter, but also the conclusions of the critique of eliminative materialism show this trend. There are not many interesting things that the contemporary debate can tell us about free will, as it is mainly based on redefining terms, at least when we consider the naturalistic part of it, as redefining terms is everything that a naturalist can do to "save" the notion of free will.

On the other hand, stances like eliminative materialism are appealing from the scientific point of view and we may not want to get rid of them entirely by replacing them with another dogmatic stance like Cartesian dualism. Hence, I presented how in Kantian transcendental idealism we can use eliminative materialism as an explanation of a person's actions when we think about them as of an appearance while preserving a meaningful notion of a morally responsible person when we think about them as of a thing in itself.

Conclusion

In this book my goal was to examine the notion of free will in the context of various stances in the philosophy of mind. Instead of answering directly the question "do we have free will?" I decided to work my way towards the notion of free will by starting with a particular concept of mind that leads to a specific anthropology and then assessing whether it is sensible to introduce a notion of free will in such a setup.

The main conclusion of my analyses is that it is impossible to come up with a sensible notion of free will in a naturalistic setup. I discussed ideas on the topic presented by Dennett and Patricia Churchland and the bottom line is that they are some ad hoc manoeuvres in order to try to preserve another, related notion, namely, the notion of responsibility. But since the ideas of free will they try to introduce do not add anything to the physicalistic picture of human beings painted by natural sciences, it is impossible to retain the notion of responsibility this way. To give a concrete argument of why I think it is so, I analysed in depth eliminative materialism in the context of artificial neural nets processing. This way, the answer to the question stated in the title should be clear – machines, as beings whose behaviour can be fully explained using the laws of nature, cannot have free will. And if human beings are like machines (e.g. similar program, but implemented in different hardware), they cannot have free will either.

The second important outcome was showing that we do not have to completely abandon naturalistic brain science to save free will and morality. Quite the contrary – eliminative materialism represented by Churchlands may be very helpful in establishing the relation between transcendental idealism proposed by Kant and the way human beings are seen by natural sciences. It is important, because in the last chapter I argue that transcendental idealism is the only way we can retain a sensible notion of free will without adhering to some kind of dogmatism, let it be naturalism or dualism. In this way I presented my solution to the third antinomy of pure reason stated by Kant in his first *Critique*. This conclusion is important because of also another reason. A person that gets to know the subject of this book may be unhappy with the fact that the most urgent question to ask, the question "do we have free will?," is not answered directly. The worst outcome could be that someone would get to a relativistic conclusion that we can choose whatever theory of mind we want and draw conclusions about free will from that. By showing the connection between naturalism on the empirical level and morality on the transcendental level, I wanted to give a strong reason to believe that transcendental idealism is a rational and the only choice to solve the problem of free will. Free will would be impossible to defend in a setup where we try to find a subject of thoughts in the empirical world.

I devoted the whole chapter to discuss the idea of free will in relation to dualism. What I tried to show is that unlike in case of naturalism, it is quite easy to establish a meaningful notion of free will on the ground of this position. The main problem of dualism lays in conceptual understanding of what is the relation between the two substances, what has been a well-known problem since Descartes introduced substance dualism.

Last, but not least, on the way to establish the above conclusions I had to refer to the most popular contemporary framework in which the problem of free will is stated – the framework that on the highest-level divides free will conceptions into compatibilist and incompatibilist ones. I argued for dropping this way of speaking about free will, as it does not introduce anything that would make solving the free will problem more easily.

Bibliography

Baker, L. R. (2004). Cognitive Suicide. In J. Heil (Ed.), *Philosophy of Mind: A Guide and Anthology* (401–416). Oxford University Press.

Buchak, L. (2013). Free Acts and Change: Why the Rollback Argument Fails. *The Philosophical Quarterly*. 63(250), 20–28.

Chalmers, D. (1997). *The Conscious Mind*. Oxford University Press.

Churchland, Paul (1981). Eliminative Materialism and the Propositional Attitudes. *Journal of Philosophy*. 78(2), 67–90.

Churchland, Paul (1993). Evaluating Our Self Conception. *Mind & Language*. 8(2), Summer 1993, 211–222.

Churchland, Paul (2013). *Matter and Consciousness*. The MIT Press.

Churchland, Patricia (2006). The Big Questions: Do We Have Free Will? *New Scientist*. 2578, 42–45.

Churchland, Patricia (2012). *Braintrust*. Princeton University Press.

Churchland, Patricia (2019). *Conscience: The Origins of Moral Intuition*. W. W. NORTON & COMPANY.

Cogley, Z. (2015). Rolling Back the Luck Problem for Libertarianism. *Journal of Cognition and Neuroethics*. 3(1), 121–137.

Corcoran, K. (2001). The Trouble with Searle's Biological Naturalism. *Erkenntnis*. 55(3), 307–324.

Davidson, D. (1973a). On the Very Idea of a Conceptual Scheme. *Proceedings and Addresses of the American Philosophical Association*. 47(1973-1974), 5–20.

Davidson, D. (1973b). Radical Interpretation. *Dialectica*. 27(3/4), 313–328.

Davidson, D. (1974). Belief and the Basis of Meaning. *Synthese*. 27(3/4), 309–323.

Davidson, D. (2001). *Essays on Actions and Events*. Oxford University Press.

Dennett, D. (1984). I Could Not Have Done Otherwise – So What? *The Journal of Philosophy*. 81(10), 553–565.

Dennett, D. (1991). *Consciousness Explained*. Bay Bay Books.

Dennett, D. (2003). *Freedom Evolves*. Penguin Books.

Dennett, D. (2017). *From Bacteria to Bach and Back. The Evolution of Minds*. W. W. Norton & Company.

Descartes, R. (1912). *A Discourse on Method*. E. P. Dutton & co.; J. M. Dent and sons.

Descartes, R. (2008). *Meditations on First Philosophy*. Oxford University Press.

Eccles, J. C. (1976). Brain and Free Will. In G. G. Globus, G. Maxwell & I. Savodnik (Eds.), *Consciousness and the Brain: A Scientific and Philosophical Inquiry* (101–122). Plenum Press.

Elzein, N. (2020). Free Will & Empirical Arguments for Epiphenomenalism. In P. Róna, L. Zsolnai (Eds.), *Agency and Causal Explanation in Economics*. Virtues and Economics, Vol. 5. Springer.

Fieser, J. (2009). *Great Issues in Philosophy*. http://springerhistory.weebly.com/uploads/2/2/0/7/22079454/great-issues.pdf

Frankfurt, H. (1969). Alternate Possibilities and Moral Responsibility. *The Journal of Philosophy*. 66(33), 829–839.

Franklin, E. (2011). Farewell to the Luck (and Mind) Argument. *Philosophical Studies: An International Journal for Philosophy in the Analytic Tradition*. 156(2), 199–300.

Hannan, B. (1993). Don't Stop Believing: The Case Against Eliminative Materialism. *Mind & Language*. 8(2), Summer 1993, 165–179.

Hume, D. (2007). *An Enquiry Concerning Human Understanding*. Oxford University Press.

Hume, D. (2009). *A Treatise of Human Nature*. The Floating Press.

Jayasekera, M. (2016). Responsibility in Descartes's Theory of Judgment. *Ergo*. 3(12), 321–347.

Jorati, J. (2017). Gottfried Leibniz [on Free Will]. In K. Tiempe, M. Griffith & N. Levy (Eds.), *The Routledge Companion to Free Will*. Routledge.

Kane, R. (1988). *The Significance of Free Will*. Oxford University Press.

Kant, I. (2004). *Prolegomena to Any Future Metaphysics That Will Be Able to Come Forward as Science*. Cambridge University Press.

Kant, I. (2007). *Critique of Pure Reason*. Palgrave Macmillan.

Kim, J. (1998). *Mind in a Physical World: An Essay on the Mind-Body Problem and Mental Causation*. The MIT Press.

Kim, J. (2008). *Physicalism, or Something Near Enough*. Princeton University Press.

Lange, M. (2008). Could the Laws of Nature Change? *Philosophy of Science*. 75(1), 69–92.

Libet, B., Wright, E. W. Jr., Feinstein, B., Pearl, D. K. (1979). Subjective Referral of the Timing for a Conscious Sensory Experience: A Functional Role for the Somatosensory Specific Projection System in Man. *Brain*. 102(1), 193–224.

Lockie, R. (2003). Transcendental Arguments Against Eliminativism. *British Journal for the Philosophy of Science*. 54 (2003), 569–589.

Lokhorst, G.-J. (2013). Descartes and the Pineal Gland. *The Stanford Encyclopedia of Philosophy*. https://plato.stanford.edu/entries/pineal-gland/

Long, A. A., Sedley, D. N. (1987). *The Hellenistic Philosophers*. Cambridge University Press.

Lycan, W. G. (2009). Giving Dualism Its Due. *Australasian Journal of Philosophy*. 87(4), 551–563.

Lycan, W. G. (2013). Is Property Dualism Better Off Than Substance Dualism? *Philosophical Studies*. 164, 533–542.

McKay, T. J., Johnson, D. (1996). *A Reconsideration of an Argument Against Compatibilism*. University of Arkansas Press.

McLaughlin, B., Bennett, K. (2018). Supervenience. In E. Zalta (Ed.), *The Stanford Encyclopedia of Philosophy* (Winter 2018 ed.). Stanford University (Winter 2018 ed.). https://plato.stanford.edu/entries/supervenience/

Moore, G. E. (2005). *Ethics*. Clarendon Press.

Murphy N. (2013). Nonreductive Physicalism. In: A. L. C. Runehov, L. Oviedo (Eds.), *Encyclopedia of Sciences and Religions*. Springer. https://doi.org/10.1007/978-1-4020-8265-8_793

Nelkin, D. (2000). Two Standpoints and the Belief in Freedom. *The Journal of Philosophy*. 97(10), 564–576.

Pitts, J. B. (2019). Conservation Laws and the Philosophy of Mind: Opening the Black Box, Finding a Mirror. *Philosophia*. 48, 673–707.

Ramsey, W., Stich, S., Garon, J. (1990). Connectionism, Eliminativism and The Future of Folk Psychology. *Philosophical Perspectives*, Vol. 4, Action Theory and Philosophy of Mind, 1990, 499–533.

Robinson, H. (2016). Dualism. https://plato.stanford.edu/entries/dualism/

Rodriguez-Pereyra, G. (2008). Descartes's Substance Dualism and His Independence Conception of Substance. *Journal of the History of Philosophy*. 46(1), 69–90.

Russell, B. (2010). *The Philosophy of Logical Atomism*. Taylor & Francis e-Library.

Schlosser, M. E. (2016). Reasons, Causes, and Chance-Incompatibilism. *Philososphia*. 45, 335–347.

Searle, J. R. (2004). *Mind: A Brief Introduction*. Oxford University Press.

Searle, J. R. (2007a). *Freedom and Neurobiology: Reflections on Free Will, Language and Political Power*. Columbia University Press.

Searle, J. R. (2007b). Neuroscience, Intentionality and Free Will. *Philosophical Explorations*. 10(1), 69–75.

Skokowski, P. (2009). Networks with Attitudes. *AI & Society*. 23, 461–470.

Stern, R. (2006). Metaphysical Dogmatism, Humean Scepticism, Kantian Criticism. *Kantian Review*. 11, 102–116.

Stoljar, D. (2015). Physicalism. In E. Zalta (Ed.), *The Stanford Encyclopedia of Philosophy* (Winter 2017 ed.). Stanford University (Winter 2017 ed.). https://plato.stanford.edu/entries/physicalism/

Swinburne, R. (2013). *Mind, Brain and Free Will*. Oxford University Press.

Van Inwagen, P. (1983). *An Essay on Free Will*. Clarendon Press.

Van Inwagen, P. (2017). *Thinking about Free Will*. Cambridge University Press.

von Wachter, D. (2006). *Why the Argument from Causal Closure Against the Existence of Immaterial Things is Bad*. https://epub.ub.uni-muenchen.de/1952/1/wachter_2006-causal-closure.pdf

Watson, R. A. (1982). What Moves the Mind: An Excursion in Cartesian Dualism. *American Philosophical Quarterly*. 19(1), 73–81.

Polish Contemporary Philosophy and Philosophical Humanities

Edited by Jan Hartman

Vol. 1 Roman Murawski: Logos and Máthēma. Studies in the Philosophy of Mathematics and History of Logic. 2011.

Vol. 2 Cezary Józef Olbromski: The Notion of *lebendige Gegenwart* as Compliance with the Temporality of the "Now". The Late Husserl's Phenomenology of Time. 2011.

Vol. 3 Jan Woleński: Essays on Logic and its Applications in Philosophy. 2011.

Vol. 4 Władysław Stróżewski: Existence, Sense and Values. Essays in Metaphysics and Phenomenology. Edited by Sebastian Kołodziejczyk. 2013.

Vol. 5 Jan Hartman: Knowledge, Being and the Human. Some of the Major Issues in Philosophy. Translated by Ben Koschalka. 2013.

Vol. 6 Roman Ingarden: Controversy over the Existence of the World. Volume I. Translated and annotated by Arthur Szylewicz. 2013.

Vol. 7 Jan Hartman: Philosophical Heuristics. Translated by Ben Koschalka. 2015.

Vol. 8 Roman Ingarden: Controversy over the Existence of the World. Volume II. Translated and annotated by Arthur Szylewicz. 2016.

Vol. 9 Tomasz Kubalica: Unmöglichkeit der Erkenntnistheorie. Leonard Nelsons Kritik an der Erkenntnistheorie unter besonderer Berücksichtigung des Neukantianismus. 2017.

Vol. 10 Renata Ziemińska: The History of Skepticism. In Search of Consistency. 2017.

Vol. 11 Jan Woleński: Logic and Its Philosophy. 2018.

Vol. 12 Wielslaw Gumula: On Property and Ownership Relations. 2018.

Vol. 13 Andrzej Zaporowski: Action, Belief, and Community. 2018.

Vol. 14 Andrzej Bator / Zbigniew Pulka (eds.): A Post-Analytical Approach to Philosophy and Theory of Law. 2019.

Vol. 15 Krzysztof Śleziński: Towards Scientific Metaphysics. Volume 1: In the Circle of the Scientific Metaphysics of Zygmunt Zawirski. Development and Comments on Zawirski's Concepts and their Philosophical Context. 2019.

Vol. 16 Krzysztof Śleziński: Towards Scientific Metaphysics. Volume 2: Benedykt Bornstein's Geometrical Logic and Modern Philosophy. A Critical Study. 2019.

Vol. 17 Jan Felicjan Terelak: Eustress and Distress: Reactivation. 2019.

Vol. 18 Roman Murawski: Lógos and Máthēma 2. Studies in the Philosophy of Logic and Mathematics. 2020.

Vol 19 Andrzej J. Noras: Geschichte des Neukantianismus. Übersetzt von Tomasz Kubalica. 2020.

Vol. 20 Tomasz Jarmużek: Tableau Methods for Propositional Logic and Term Logic. Translated by Sławomir Jaskólski. 2020.

Vol. 21 Marta Kudelska: Why Is There I Rather Than It? Ontology of the Subject in the Upaniṣads. 2021.

Studies in Philosophy, History of Ideas and Modern Societies

Edited by Jan Hartman

Vol. 22 Georg Schmid: The Treachery of the Elites. On Political Discontent. 2021.

Vol. 23 Hanna Urbańska: The Philosophical System of *Śiva Śatakam* and Other *Śaiva* Poems by Nārāyaṇa Guru. In Relation to *Tirumandiram* by Tirumūlar. 2022.

Vol. 24 Marcin Krasnodębski: Green Chemistry. A Brief Historical Critique. 2022.

Vol. 25 Artur Przybysławski: Thales and the Beginnings of European Reflection. 2023.

Vol. 26 Krzysztof Krenc: Can Machines Have Free Will? The Concept of Free Will in Relation to the Psychophysical Problem. 2023.

www.peterlang.com